Lecture Notes in Intelligent Transportation and Infrastructure

The series “Lecture Notes in Intelligent Transportation and Infrastructure” (LNITI) publishes new developments and advances in the various areas of intelligent transportation and infrastructure. The intent is to cover the theory, applications, and perspectives on the state-of-the-art and future developments relevant to topics such as intelligent transportation systems, smart mobility, urban logistics, smart grids, critical infrastructure, smart architecture, smart citizens, intelligent governance, smart architecture and construction design, as well as green and sustainable urban structures. The series contains monographs, conference proceedings, edited volumes, lecture notes and textbooks. Of particular value to both the contributors and the readership are the short publication timeframe and the world-wide distribution, which enable wide and rapid dissemination of high-quality research output.

More information about this series at http://www.springer.com/series/15991

Roman Gumzej

Intelligent Logistics Systems for Smart Cities and Communities

Springer

Roman Gumzej
Faculty of Logistics
University of Maribor
Celje, Slovenia

ISSN 2523-3440 ISSN 2523-3459 (electronic)
Lecture Notes in Intelligent Transportation and Infrastructure
ISBN 978-3-030-81202-7 ISBN 978-3-030-81203-4 (eBook)
https://doi.org/10.1007/978-3-030-81203-4

This Springer imprint is published by the registered company Springer Nature Switzerland AG
The registered company address is: Gewerbestrasse 11, 6330 Cham, Switzerland

To Stane, Vali and Eva.

Foreword

Many scholars work on technology use and how innovative technology can enhance logistics and service operations. Few researchers blend pragmatism, vision, and a sense of realism in the way that Dr. Gumzej does in his work. Innovation requires more than just a "new great thing"—it requires the ability to change the world because it is practical and applicable, solving genuine problems and, therefore, providing value and solutions to professionals. The way that Dr. Gumzej approaches the problems that he studies ensures professional interest in the solutions while also easing and reducing the problems that we face in our societies. I have had the pleasure of meeting Dr. Gumzej through my role as Editor-in-Chief for the *International Journal of Applied Logistics*, and learning more about his work and the work his team undertakes. His style of research is an excellent fit with our journal's focus, and I always look forward to seeing a submission from him or his team. My research work has focused more narrowly on logistics and supply chain applications, but always with a healthy dose of technological change and improvement. My emphasis on developing a better world through the adoption and use of technology also influences my perspectives in another journal I edit, the *International Journal of Sociotechnology and Knowledge Development*, and in my role as an Associate Editor in the *Journal of Supply Chain Management*. From my work with researchers around the world, I am acutely aware of the significant changes and improvements that our societies can derive from innovations and the use of new technologies—if the research is not locked away in academic- or research-oriented forums. This volume opens up the research to a wide audience that will benefit from the knowledge and expertise.

This volume of Dr. Gumzej's work draws together a collection of his most exceptional research. Together, they show a pathway for a better tomorrow, moving from discrete technologies (e.g., the use of smart devices in Chap. 2) through to wider changes (e.g., Industry 4.0 applications in Chap. 7, leading to Logistics 4.0 in Chap. 8). It focuses on providing support for change in practice and use of innovations in Chaps. 13–20, where the emphasis is on the use-cases of the technologies. In this way, the volume has benefit both to researchers and students in this area (in the first half of the book), as well as professionals and business people (in the second half of the book) who seek to understand how to unlock value in their industry and business.

I appreciate the flow and development of the book. It leads readers through a gentle progression of concepts, technologies, and ideas. In this way, it can be used as an introductory reader with more advanced students skipping some earlier chapters and instead turning to the detailed chapters where Dr. Gumzej goes deeper into the applications of technologies. In all cases, care has been taken to present the ideas and models that inform the application and use of technologies in those circumstances, providing a clear framework and support for the readers. I am particularly fond of the use-cases near the end of the volume. This is important, in my mind, both for students studying the topics to see and understand how to go from an abstract discussion of technology through to the implementation and use. It also gives value to the users and professionals that may look to adopt these approaches, providing insight into the implementation and use and providing a vision for how the technology can unlock value in their organisation and solve their problems.

As the volume brings together original ideas, it can be valuable in a range of different classes, such as logistics, healthcare, and information systems or information science classes. This multi-disciplinary approach is one strength of the volume and shows how the principles can provide multiple sources of value to different organisations with often different needs.

Having closely followed some of Dr. Gumzej's research in my editorial capacity, I am delighted to see this volume draw together his exceptional work in one place to benefit readers. Whether the reader is looking to progress their study or enhance value in their business through innovation, this book will provide insight and enhance understanding. The ability to present the body of work in this way allows readers to maximise their exposure to the valuable work Dr. Gumzej has undertaken and benefit from the range and breadth of the innovation applications. I am excited about the book and the insights it offers, the vision for our future, and will be extremely pleased to recommend this to my future students and colleagues. This book belongs on the shelf of any business professional interested in these topics and many students in these areas; I expect it will quickly become a classic volume and will be one of the most read books on the shelves of those who have a copy.

Associate Professor Lincoln C. Wood

Department of Management, Otago Business School, University of Otago (New Zealand).

Adjunct Research Fellow, School of Management, Curtin Business School, Curtin University (Western Australia, Australia).

Editor-in-Chief—*International Journal of Applied Logistics*

Editor-in-Chief—*International Journal Sociotechnology and Knowledge Development*
Associate Editor—*Journal of Supply Chain Management*

June 2021

Lincoln C. Wood
University of Otago
Dunedin, New Zealand

Preface

The objective of this book is to stress the synergistic influence of logistics and informatics on the future of our information society. Since the emergence of the Internet, information technology has played the enabling role in two key fields of logistics—supply chain management as well as transport and traffic management—with the goal to ensure their timely, correct and dependable operation, and to provide for sustainable growth of the world's economy. Sustainable and smart are terms, usually written out in capital letters these days, characterizing anything future-oriented.

As logistics has made industry re-think their value chain, information technology has brought society sufficient information resources to enable it to grow its knowledge and gain from it. To support smart production, transportation and living, new information infrastructure has been established. Broadband networks have been laid out, smart sensors have been installed, global navigation and tracking services have been introduced, etc. Information systems and their services are becoming more and more intertwined. Systems are no longer made of components, but may also comprise systems—systems of systems. Computer networks are transferring data among interconnected networks of networks. Information systems are becoming autonomous systems only requiring high-level instructions to perform their functions. Knowledge bases, warehouses, and markets are superseding databases. Inevitably, this leads to the realization of the slogan, too often misused before, that "knowledge is power". Nowadays, not only everyone can access it, anyone can also contribute to it, and do so anytime and from almost any place on earth. All knowledge, previously found only in certain places like historic sites, libraries, and government archives, is becoming available to all citizens. Hereby, we—the digital citizens—are being fully integrated into the information society. By the process of "digitization", everything physical is becoming its digital representation. By a mouse-click, touch on a mobile terminal's screen, or even by simply looking at an object through smart glasses, one can instantly get useful information about a person, a building, a store, a product, a service, etc. Based on information received, one makes informed decisions and takes actions, which do not only take place in the physical world but also in the cyberspace. Every new smart device and service raises new opportunities, questions and issues related to its application, steering its evolution towards a more efficient and useful entity in the information society. By striving to do things smart, sustainable and green logistics inevitably gets involved. By the integration of smart devices into systems with integrated logistics support, intelligent logistics systems being considered the enablers of sustainable growth and the arising innovation society are being devised.

In this book, the major challenges of the new innovation society's information infrastructure are discussed. To start with, in the first chapter the terms linking smart cities and communities with smart devices and services are introduced. In the following two chapters, the underlying information technologies and services are described and classified. When speaking about the new intelligent systems and services of the intelligent web, we are immediately confronted with issues concerning information systems' safety and security. Since our lives are intertwined with matters of dependability and privacy, chapter four is devoted to the methods and mechanisms for ensuring safety and security in the globally digitized world of smart things. In the following chapters, the different types of smart systems and services influencing our daily lives are discussed and exemplified by use cases reaching from e-health and telemedicine over e-commerce, intelligent production and transportation to smart cities and communities. In the conclusions, the influence of the evolution of smart systems and services on our environment as well as the arising innovation society of the future are summarized.

As the title implies, this book sets the modern infrastructure of smart devices and services into the perspective of the future smart cities and communities. In the course of this, it discusses the major technological solutions and steps towards integrated logistics solutions to be used in these environments with their benefits in terms of efficiency, interoperability, and sustainability. By doing so it paves the logistician's way towards the aspired innovation society.

Maribor, Slovenia
February 2021

Roman Gumzej

Contents

Chapter 1
Introduction

...You may say, I'm a dreamer, but I'm not the only one...

John Lennon

In times of the Internet, global communication, navigation and information accessibility have become a matter of course. They are the corner stones of the *Information Society* (*Society 4.0*, Fig. 1.1) we live in, since they provide for the right information at the right times in the right forms and thereby help us make the right decisions. Technological advances have been made in the form of incorporating information technologies in the form of *smart devices*, ranging from appliances to industrial machinery and all sorts of transportation means.

With industrial development we strive for *intelligent products*, ones that shall be "better" in all respects and won't become sources of pollution once their *life-cycles* are over. To achieve this goal, *smart production* is necessary to suit the diversity of our needs as well as to keep its footprint to a minimum. Since logistics plays a major role in supply and distribution as well as the orchestration of production, next generations of intra- as well as inter-organizational logistics are being introduced as parts of the *Industry 4.0* and *Intra-/Interlogistics 4.0* initiatives.

A *Smart City* needs to offer its dwellers a healthy and inspiring habitat in which they can fulfill their goals and, since their lives would be less consumed by labor intensive tasks, they could be more fulfilled with innovation. *Innovation* is expected to become the dominant "industry" of the future and will hopefully steer our lives towards a better future for ourselves and the future generations.

As the *digital transformation* increasingly impacts the modern society, there are many unanswered questions about how to proceed. The Internet's position as a worldwide entity requires global answers to these questions. Unlike issues with conventional infrastructure, which must be addressed locally, the internet infrastructure is truly global, and in fact in many ways extra-terrestrial, which demands for comprehensive, inter-operable solutions. Several examples of such issues include the con-

R. Gumzej, *Intelligent Logistics Systems for Smart Cities and Communities*,
Lecture Notes in Intelligent Transportation and Infrastructure,
https://doi.org/10.1007/978-3-030-81203-4_1

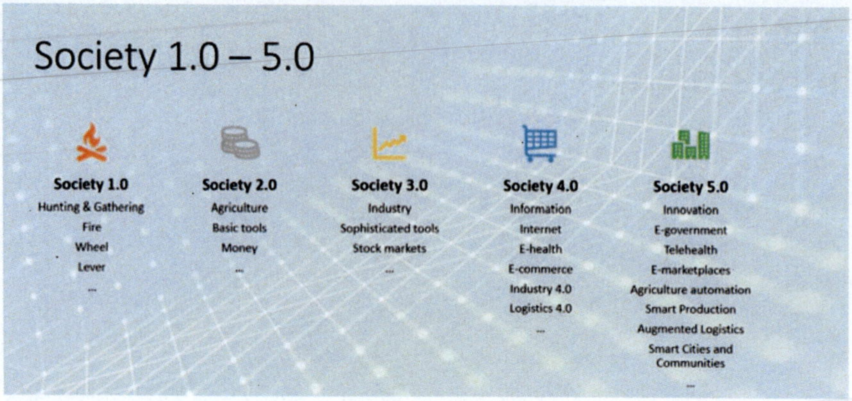

Fig. 1.1 Society 1.0–5.0

cepts of *global communication*, *global timekeeping* and *global navigation*, together with new unified and more precise definitions of the established *measurement units* to provide for increased volume or precision data as well as interoperability of smart devices and services. This should result in a new *industrial revolution*, since all industries are expected to make a quantum leap into a new era, hopefully characterized by an improved overall *quality of life* and *product quality* in terms of the *0-emission policy*.

People are mostly aware of the great impact that the *Internet*, *World Wide Web* and associated *information technologies* (IT) have on our society's social, economical and environmental aspects. The *Web* is so intertwined that it became one of the cornerstones of our professional, scientific as well as private activities. As a tool for *advanced information acquisition*, *data mining* and *machine learning*, the current Web has just about reached its boundaries. Hence, a new *Web 2.0*, also called the *Intelligent Web* or *Wisdom Web*, is slowly but surely becoming a reality. It shall incorporate several novel *information classification* capabilities and different *Knowledge Management* services – hence, it is termed the Intelligent Web.

In order to integrate devices and services into the Intelligent Web, their *digital twins* are formed, as part of the *digitization* process, associating objects in the Web with physical objects, providing information about their properties and abilities.

Every Web 2.0 object shall be aware of itself and other objects co-existing in the *hyper world* to provide for *4Rs of information systems*, providing:

1. the right information/*service*, to
2. the right person/*entity*,
3. at the right time, and
4. in the right form *and context*.

as well as *7Rs of logistics*, getting:

1. the right product/*service*,
2. in the right quantity/*amount*,
3. in the right condition,
4. at the right place,
5. at the right time,
6. to the right customer/*entity*, and
7. at the right price.

Several noticeable differences, indicating the paradigm shift, have been emphasized in *italics*. The orientation has moved from people towards objects, representing the entities of the Web 2.0. At the same time the context has become equally important as the form of information, since the focus has changed and is now considered ambivalent, considering the involved entities. Also, service is considered a product, since the products are abstracted and coexist with them within the same space and time. Hence, in terms of smart production and logistics, we now speak about the *Internet of Things* (IoT) and *intelligent agents* managing and promoting them through the arising *Physical Internet* (PhI).

Obviously, these goals are not achievable without solving some infrastructural issues first. *Green technologies*, *intelligent materials*, *ubiquitous computing*, *global communication*, *global navigation/tracking* and *global time* services represent the basic information infrastructure of the *Information Society*. Together with the devices and services of the Intelligent Web they need to be sustainable and inter-operable in order to last and grow. Hence, they require *Integrated Logistics Support* incorporated in the basic functionality of *Intelligent Logistics Systems*. They may act as or employ agents as software solutions that are capable of helping their users in getting things done more efficiently by autonomously making better, more informed decisions, self-improving when executing prescribed tasks within an ever changing environment. They make use of *intelligent agent technology* and the *Intelligent Web services* of the Web 2.0. As Wisdom-Web-oriented personal/business assistants the *ubiquitous agent communities* will on one hand recognize the user-intent and then offer best possible solutions through dedicated services to their users.

Eventually, one may ask oneself: "Is the intelligent Web ready?" In other words, can decision making be justifiably placed into the hands of agents or can their decisions mislead or even bring harm to their users? How about safety and security in a digital world? How can one foresee the changes and their affect on the arising *Industry 4.0*, *Logistics 4.0*, and *Innovation Society* (*Society 5.0*). It is the purpose of this book to give an overview and provide some crucial answers about the information technologies that shall greatly affect our society in the future and highlight the role of logistics as the carrier of the next "industrial" revolution.

Chapter 2
Smart Devices

Smart devices should be able to interconnect among each other and cooperate with other such devices in their proximity to accomplish their goals. Depending on the nature of these devices and physical systems they are connected with, they are associated with various levels of risk. Hence, appropriate security measures need to be implemented to prevent damage from their failure, malfunction or even misuse. To achieve this, not only they must be self-aware and self-reliant in their operation, making them "smart". In addition, their safety and security must be provided for locally as well as globally within the IoT.

2.1 Introduction

Nowadays, we are often speaking of *smart devices* in the context of the *Internet of Things* (*IoT*). What we actually mean, is that these interconnected devices can as such co-operate in the *hyper world*. Besides their ability to be monitored and controlled on the Internet, their life-cycles can easily be managed by their producers in order to achieve their *0-emission policy*. Due to the global nature of the IoT, they rely on the presence of the global communication, timekeeping and navigation infrastructure.

Smart devices range from simple sensors and actuators, known as components of the *Supervisory Control And Data Acquisition* (*SCADA*) systems, to more complex control systems or even entire plants, being accessible and controllable from a remote location through the Internet (Fig. 2.1). On the other hand smart devices also comprise *smart utility devices*, which better fulfill their functions by their (inter-)connectivity. In this chapter an overview of the smart technologies is given and an outlook of their future developments is laid out.

R. Gumzej, *Intelligent Logistics Systems for Smart Cities and Communities*,
Lecture Notes in Intelligent Transportation and Infrastructure,
https://doi.org/10.1007/978-3-030-81203-4_2

Fig. 2.1 IoT interconnects smart devices and their services

2.2 Global Communication

Since Internet access is primarily driven by market forces, global communication in sparsely populated parts of the world is still not considered a matter of course. The integration of worldwide population by an appropriate communication network has gone different ways, from upgrading nation-wide *Public Switched Telephone Networks* (*PSTN*) with the *Global System for Mobile communications* (*GSM*) to extending the existing microwave radio-communications-based telecommunication to satellite communication systems (e.g. Iridium).

Although communication networks were originally meant for voice communications only, their development led to audio-visual and data communication in the broadest sense, especially with the digitization of transmitted data and their transfer by packet switching through the Internet Protocols. Copper-wires were mostly replaced by higher capacity as well as more reliable and lower-noise lines, like coaxial and optical fibers. Microwave communications were upgraded by laser communications wherever possible to increase transfer speeds and communication channels capacity.

In sparsely populated areas, neither of the aforementioned communication technologies found economic justification. While in cities with a high population density and broadband demands it is economically viable to establish high-cost optical fiber connections, in sparsely populated areas outside of cities, other solutions are being employed to satisfy both customer demand and profits of the shareholders. To bring the Internet to more remote areas, in addition to costly satellite communication, GSM base-stations as well as microwave wireless routers have been introduced as an extension to PSTN networks, acting as replicated public Internet hot-spots. Here, a popular option is the use of *Worldwide Interoperability for Microwave Access* (*WiMAX*) technology [1]. WiMAX is a point-to-multi-point system, running the *IEEE 802.11 protocol*, that operates on a similar base-station principle like the GSM networks. It allows for broadband access in sparsely populated areas by utilizing WiMAX hot-spots as access points transmitting data to customer's receiving antennae via radio-communication. WiMAX can transmit over numerous radio frequencies, which not only avoids interference with other radio-signal based transmissions, but also allows for different types of data to be transmitted simultaneously according to their different transmission speed and band requirements. This system is easy to build and maintain, and is thus considered cost-effective.

Nevertheless, neither WiMAX nor any other microwave technology is not without problems. Since radio frequencies are finite and a great amount of the radio spectrum has already been reserved both by government and private organizations for various purposes, their choice is limited. Lower radio frequencies, which are able to penetrate solid objects, are more in demand and harder to come by. If, however, one can allocate an appropriate spectrum to it, WiMAX broadband access can represent a viable telecommunication solution, even in the most secluded parts of the world.

GSM networks have reached a maturity level to a degree where they no longer serve as a backup communication network to wired communication technologies. Already in the 4th generation of global mobile communications (*4G*) networks they have achieved transmission speeds by which they can compete with public and most private WiFi networks. In their 5th generation (*5G*) they tend to make a quantum leap forwards by a drastic improvement of transmission speeds and the use of latest internet protocols (*IPv6* with the *Neighbor Discovery Protocol* (*NDP*)) that shall make them the enabling technology of the IoT.

2.3 Global Timekeeping and Navigation

Another technological development facing challenges from the lack of available carrier frequencies is the *Global Navigation Satellite System* (*GNSS*). It consists of multiple constellations of navigation satellites, e.g. *GPS* (USA), *GLONASS* (Russia), *Galileo* (Europe), *BeiDou* (China), *QZSS* (Japan), etc. While some share common frequencies (e.g. QZSS and GPS), most have their own carrier frequencies across which they transmit their exact position and time to receivers that determine their own location and time based on this data. Due to its pioneer role, the name of the first *Global Positioning System* (*GPS*) is often used interchangeably with the GNSS, however the GPS network of satellites nowadays is only one of several that serve the same purpose (Fig. 2.2).

According to [2], locations on earth can be determined using some 24 satellites orbiting the Earth in six orbital planes every 12 h at an altitude of around 20.200 km.

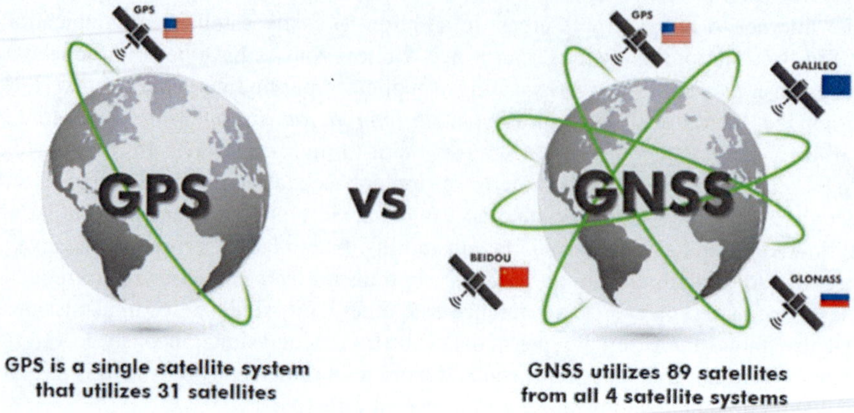

Fig. 2.2 GNSS constellations of satellites that orbit the Earth

Each satellite carries an on-board atomic clock, which allows it to transmit extremely accurate time-coded signals to a GNSS receiver. The receiver then calculates the distance to the satellites based on the times that were necessary to receive their respective signals. A common characteristic of all GNSS is that at any given time from any location on earth at least 4 need to offer a good enough reception to enable such navigation. Many receivers noways are able to use data from multiple GNSS constellations, which increases their number of "visible" satellites and enhances the precision of their calculated location.

Besides communication, time synchronization has become a pressing issue in the globalized world. Due to large numbers of international transactions—passenger commutes, cargo traffic, stock moves, monetary transfers, …—that need to be monitored in real-time, possibly across multiple time zones, local time-keeping based only on the accuracy of the clock mechanisms and their maintenance is not an option anymore. Radio clocks and watches have been very popular in Europe since the late 1980s and, in mainland Europe, most of them use the DCF77 signal to set their time automatically. Further industrial time-keeping systems at railway stations, in the field of telecommunication and information technology, at radio and TV stations are radio-controlled by DCF77 as well as tariff change-over clocks of energy supply companies and clocks in traffic-light facilities.

DCF77 is a German long-wave time signal and standard-frequency radio station. The highly accurate 77.5 kHz (approximately 3868.3 m wavelength) carrier signal is generated from local atomic clocks that are linked with the German master clocks at the PTB in Braunschweig. The timestamp sent is either in Coordinated Universal Time (UTC)+1 or UTC+2 depending on daylight saving time. Being used as the primary source of exact time in Europe from 1959, similar infrastructure has grown over its borders to cover most of the earths surface by implementing additional transmitters operated by local UTC reference laboratories in Russia, China, Japan, Great Britain, and USA.

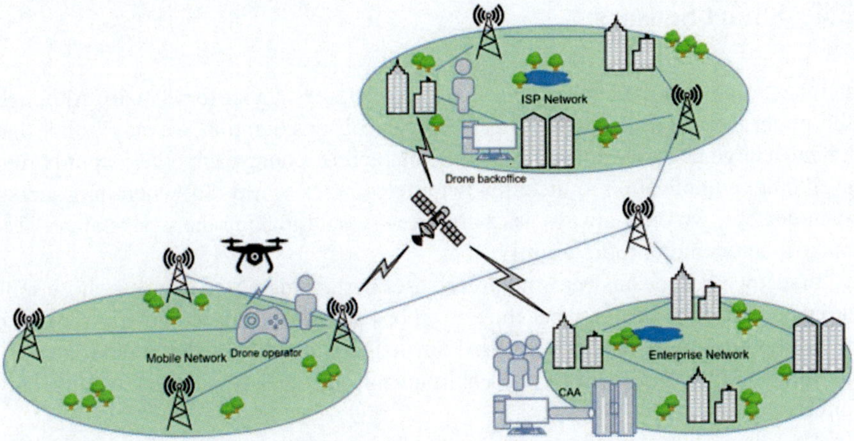

Fig. 2.3 Drone's, sports-aircraft's and helicopters operating environment

Together with the Coordinated Universal Time (UTC) reference laboratories and a network of Longwave, Shortwave, VHF/FM/UHF, RDS CT, and Satellite (GNSS) transmitters the Internet has become the most accessible and reliable source of accurate time on Earth. The Internet time synchronization protocols (Network Time Protocol (NTP) and Precision Time Protocol (PTP)) are used to transmit exact UTC (GMT) time across the Internet to citizens and companies, alike. Based on accurate time and ephemeris information, contemporary satellite navigation can provide for location accuracy of up to 1m, which brings unprecedented navigation and tracking possibilities into transport and logistics. On the other hand the notion of the correct time of events provides validity to transactions, so one not only knows where something occurred, but also when, and whether this was on time.

To summarize the above considerations, let's look at an example. In order to operate a global network of IoT devices one must consider global communication in conjunction with global navigation and timekeeping infrastructures. *Drones*, for example (Fig. 2.3) rely on all these infrastructures to fulfill their missions on track/time and return safely to their points of origin. Their missions start in the field and may also be time-triggered. Their oversight must be assured to ensure flight safety. They may be controlled by a drone operator on site or from the backoffice. The backoffice needs to correspond with the *Civil Aviation Authority* (*CAA*) to properly plan and fulfill drone flights, thus allowing for their flight routes to be accounted for by the CAA's air-traffic management operations. Only then the security of air-traffic can be assured also in the class G airspace, which drones share with other small air-crafts, according to the shared standards on air-traffic safety.

2.4 Smart Sensors

In today's age of smart phones, one would expect every device to be smart. Although this might seem logical at first glance, after careful consideration, we may realize that having a large diversity of smart devices, interacting among each other, would bring even more complication to the already quite complex world. So, when introducing such devices, we must always ask ourselves—would making the device smart also make it better and/or more useful?

In many cases the answer is no. Such cases pertain mainly to devices whose self-reliance defines them for what they are. For example, using an electric tool, like a flint-borer entirely depends on the knowledge and skills of the person using it. Its function is not affected by the environment, nor does it need to exchange any information with it.

The main characteristic of the devices, where the answer is yes, is that they are better in fulfilling their function in conjunction with each other and possibly other related devices. They may also better fulfill their purpose when connected to a knowledge base or contribute their services as a part of a larger set-up. On a small scale, smart warehouses, factories and homes are examples of such systems. On a large scale, however, we may look at smart power grids, cities and the physical internet as examples of such systems. In these cases, utilizing smart sensors, actuators and control systems makes them better, more effective, easier to maintain and also more transparent, from user as well as control system perspective.

So, what are smart sensors? *Smart sensors* are sensing devices with integrated transceivers for communication among each other and/or their control systems. They may be as simple as a temperature sensor or as complex as a *body sensor network*. Either way, they communicate their sensor data to the back-end information system and may also be controlled by it. In some cases intelligent devices can cooperate among each other, e.g. RFID Class 5 devices with proximity sensors. In the sequel some of the aforementioned and some additional sensors are classified and described in more detail.

2.4.1 Simple Sensors

Simple sensors detect (changes in) environmental conditions and report their status as read-outs to a back-end information system (e.g. Fig. 2.4). They may be directly connected to their control systems or rely on various communication networks (e.g. Fieldbus, Industrial Ethernet, (W)LAN, (W)WAN, Bluetooth, Z-wave, etc.) to do so. The values they report are either binary (presence, shutter control, smoke detectors), quantitative (temperature, wind, power consumption), or qualitative (brightness, water level, CO/CO_2 detectors).

Their reported states are processed by back-end information systems, providing for monitoring of their read-outs as control actions, such as issuing alarm signals

Fig. 2.4 Simple light sensor for the https://www.arduino.cc/Arduino; photo by https://www.arduino.cc/Arduino, distributed under CC https://creativecommons.org/licenses/by-sa/3.0 BY-SA 3.0 license

Fig. 2.5 IR proximity sensor for the https://www.arduino.cc/Arduino; photo by https://www.arduino.cc/Arduino, distributed under CC https://creativecommons.org/licenses/by-sa/3.0 BY-SA 3.0 license

and/or sending signals to actuators to perform some regulatory actions. Since, simple sensors cannot perform any actions by themselves, some oversight across the sensor-states is necessary. Reactions are usually triggered by signals to actuators or alarms in cases of emergency states (e.g. water spill, high CO concentration detection), reported to the back-end system, some erratic behavior of the control system or sensors (e.g. false alarms or values outside thresholds), or blackouts due to network or sensor failures.

2.4.2 Proximity Sensors

Proximity sensors are sensors that are able to detect the presence of nearby objects without physical contact (e.g. Fig. 2.5). Examples of proximity sensors are motion detectors, parking sensors (Doppler, photoelectric, ultrasonic sensors), identification sensors (e.g. RFID, NFC), etc.

They are often used for state monitoring and as parts of smart devices to monitor the state of their operation environment in order to be able to take some more informed or coordinated action.

Fig. 2.6 Air quality sensor for the https://www.arduino.cc/Arduino; photo by https://www.arduino.cc/Arduino, distributed under CC https://creativecommons.org/licenses/by-sa/3.0 BY-SA 3.0 license

2.4.3 *Sensors-Actuators*

Sensors-actuators are able to act according to their condition or in response to a triggered action from the back-end information system.

For example, if a proximity sensor reports a presence in the room and the brightness sensor (Fig. 2.4) reads "too dark", a smart switch may turn on the lights and starts transmitting energy consumption values. Based on the description, only the last of the aforementioned three sensors actually is a sensor-actuator, although the former two could be (and in many cases are) parts of it.

Similarly, if a CO concentration sensor (e.g. Fig. 2.6) reads "too high", a smart switch may turn on an alarm and a warning sign. As a remedy action, this may be followed to an activation signal to an active ventilation system to reduce CO concentration. In the sequel the control system would monitor CO sensor status read-outs and switch off active ventilation, once it's state returns to normal again (according to CO sensor calibration).

The level of sensor integration depends on how "smart" a sensor is—the smarter the sensor the less control from the back-end system is required.

2.5 Nanotechnology

As the scale of information systems grows, the size of their components is being reduced. In the future, sensors and systems on a nanoscale are expected to spread. Unlike "conventional" electronic devices they represent an entirely new technological branch. Hence, the term nanotechnology. Nanotechnology is so small, it's invisible to the human eye, but can generate extraordinary amounts of data. It can also produce new smart nanomaterials that can harness our environment for energy, security, sustainable food production, healthcare, education, media, digital finance, big data, transport and water resources.

Nanotechnology represents science, engineering, and technology conducted at the nanoscale, which is about 1–100 nm [3]. Nanotechnology and nanoscience study the behavior and application of extremely small objects and can be used across all other science domains, such as chemistry, biology, medicine, physics, materials science, and engineering. Nanotechnology is not just a new field of science and engineering, but also a new way of looking at and studying them. According to [4] the semiconductor industry had already entered the nanotechnology world in 2000 with the introduction of the 130 nm node with a 70 nm length gate, followed by the 90nm node featuring a critical dimension gate of 50 nm in 2002. In future green smart cities, one can imagine that:

- bulk band structure of solids will be replaced by geometry—dependent energetic structures of nanostructures,
- doping, bulk processes are going to be replaced by the precise manipulation and placement of individual atoms,
- crystal growth—bulk processes are going to be replaced by self-organization of matter and self-assembly of complex structures.

Nanotechnology in future green *Smart Cities* [5] will expand to include molecular nanosystems—heterogeneous networks in which molecules and supra-molecular structures serve as distinct devices. The proteins inside cells work together in this way, but whereas biological systems are water-based and markedly temperature-sensitive, these molecular nanosystems shall be able to operate in a far wider range of environments and should be much faster.

With the development and introduction of nanotechnologies in the physical world unprecedented application opportunities shall arise. Computers and robots will be reduced to extraordinarily small sizes. Medical applications as ambitious as new types of genetic therapies and anti-aging treatments would become conceivable and more accessible. New *Human Machine Interfaces* (*HMI*) linking peoples minds directly to electronic devices would take ubiquitous computing and augmented reality to a new level. Any industry and health care domain can benefit from the nanotechnology of the future. It might also aid the preservation of the environment by more efficient resource usage and better waste management methods [6].

The future nanotechnology will very likely include the use of nanorobotics. Nanorobots would take on human tasks as well as tasks that humans could never complete. For example, the depleted ozone layer could be renowned by nanorobots that could single out molecules of water contaminants and aid its restoration, etc.

Another area where nanotechnology will augment the infrastructure of smart cities is through the *Internet of Nano Things* (IoNT). The IoNT is a similar concept to the IoT, in that it uses sensor networks and hubs to collate the data, routers to send the data over long distances, and advanced software methods, but a the nanoscale. Whilst a subset of the huge volumes of data that is possible with IoT data networks, it will manifest itself as a dynamic tool for remote and environmental monitoring applications, as well as in personalized telemedicine, medical devices and substantially aid preventive healthcare.

The healthcare in future smart cities should improve using nanotechnology. The revolution in telehealthcare shall bring advanced point-of-care diagnostic tools to our home environments as well as much more accurate self-diagnosis using highly advanced nano-enabled mobile devices to track our health in real time by mobile health applications. Physicians shall be able to detect an illness or injury early and attack it at the molecular level by nanosurgery. Among others, this could eradicate cancer, as the "surgical" procedures would be done with utmost precision on the cellular level. Cancer cells would be identified, removed, and subsequently replaced by healthy cells grown from stem cells.

For the time being nanotechnology is still in a very experimental phase. Its main problems originate from the size of its enabling technology and unpredictability associated with its engineering. This pushes its aspired goals and solutions into a somewhat distant future, but its concepts are already well established awaiting technological maturity.

2.6 Metric Systems

To get smart devices to interact and seamlessly work together, standards are needed. For example, while the physics of the GNSS system are fairly straightforward and universally accepted, there are some fundamental questions, which have never been finally resolved, like: "What is time and how do we measure it? What is a second?"

Officially, according to the 13th General Conference of Weights and Measures (Resolution 1, 1967) a *second* is "the duration of 9,192,631,770 periods of the radiation corresponding to the transition between the two hyper-fine levels of the ground state of the cesium-133 atom" [7]. An internationally accepted reference for time is called *Coordinated Universal Time* (*UTC*), which is calculated using approximately 230 clocks from 60 laboratories in different locations globally in order to take an average reading, which soothes out measurement anomalies created by internal and external deviations.

Increasing sophistication and accuracy required to operate the GNSS and other digital platforms will require a precise, always accessible universal time source. The principal drawback of UTC is that it cannot be accessed in real-time. Therefore, perfect synchronization, needed by the telecommunication networks, is not yet possible. The new atomic clocks, based on laser spectroscopy of a single trapped ion, represent a possible solution. They are 100 times more accurate than current atomic clocks. In fact, they are so accurate that, if such a timepiece were in use since the universe began, it would now be off by only about 30 s (SuperGPS).

As pointed out in [8], just as time's definition must be agreed upon, so too must we establish a common definition of metric prefixes for the computer industry. The logic of the metric system is clearly based on factors of ten, and prefixes are given accordingly. There is no dispute over what a kilogram or a deciliter is. The world of computers does not, however, work on powers of ten. Because they are based on binary logic, the digits inside computers are combinations of the numbers 2, 4, 8, 16,

32, 64, 128, and so on, increasing by powers of 2. So, what does it mean, when we refer to a kilobyte? Metric logic would have us believe this is 1,000 bytes. Binary logic, on the other hand, reveals that a kilobyte is actually 1,024 bytes. One solution, also approved by the International Electrotechnical Commission, is an entirely new system of metric prefixes to reflect numbers based on powers of 2 as opposed to 10. The introduction of terms such as mebi-bytes, tebi-bytes, and exbi-bytes would accurately describe binary values, but would they lead to clarification or confusion? For the vast majority of the world's computer users, it is not important whether a 1-gigabyte hard-drive has a memory size of 1,000,000,000 or 1,073,741,824 bytes, but in the future it may be.

As with the question of what is a second, the necessities of the ever increasing complexity of smart devices and their integrations into larger setups like "systems of systems" shall require universally understood standards. Since these terms and measures have become generally accepted, one might consider this question obsolete, however considering a possible paradigm shift in connection with nanotechnologies and quantum computing let one wonder. With the advent of multi-state gates, where a single *qbit* of information may have three or more states, the base of the numbering in computing would change appropriately. Considering the time it took to devise a viable definition of a second, let one wonder, when these quantities would render universally accepted standard measures that would be equally understood and accepted globally.

2.7 Synopsis

Smart environments, wearable computers, and ubiquitous computing in general represent the upcoming *sixth generation* of computing and information technology. These devices will be everywhere – clothes, home, car and office—and their economic impact and cultural significance will dwarf those of the former generations. Smart cities are expected to possess audio-visual and haptic interfaces to people's environments like apartments, cars, and offices.

Much as a mobile phone has become an essential part of our modern lives, we are rapidly approaching a time when it will be unthinkable to drive away without audio-visual interaction with our GNSS system, enter a dark room without looking for a switch to turn on the lights, control our home appliances by voice, receive notifications about important events throughout the day and work in a clean, self-regulating environment. Few people, however, give much thought as to how this could be accomplished.

According to [9], these applications will give machines perceptual abilities that will allow them to interact with people naturally—to recognize people and remember their preferences and peculiarities, to know what they are looking at, and to interpret their words, gestures, and unconscious cues, such as vocal prosody and body language. Facial-expression recognition shall be able to interact with other smart-environment capabilities. People shall use various types of computers, cameras, microphones and other sensors that may be worn, handheld or integrated into their clothing (cp. Chap. 13). These wearable devices shall adapt to a specific user and be intimately and actively involved in the user's activities. For example, people might have built in cameras into smart eyeglasses or lenses with face recognition software that would be able to help them remember the name and other details of the person, they are interacting with, by whispering it into their ears.

References

1. Vaughan-Nichols SJ (2004) Achieving wireless broadband with wimax. Computer 37:10–13. https://doi.ieeecomputersociety.org/, https://doi.org/10.1109/MC.2004.4
2. Bretz EA (2000) X marks the spot, maybe [gps navigation]. IEEE Spectr 37(4):26–36. https://doi.org/10.1109/6.833025
3. Nano.gov: What is nanotechnology? https://www.nano.gov/nanotech-101/what/definition
4. Bourianoff G (2003) The future of nanocomputing. Computer 36(8):44–53. https://doi.org/10.1109/MC.2003.1220581
5. Roco MC Nanotechnology's future. https://www.scientificamerican.com/article/nanotechnologys-future/
6. Nanogloss.com: The future of nanotechnology. https://nanogloss.com/nanotechnology/the-future-of-nanotechnology/
7. Allan DW, Ashby N, Hodge C (1998) Fine-tuning time in the space age. IEEE Spectr 35(3):42–51. https://doi.org/10.1109/MSPEC.1998.663757
8. Mcfedries P (2013) Tracking the quantified self [technically speaking]. IEEE Spectr 50(8):24–24. https://doi.org/10.1109/MSPEC.2013.6565555
9. Pentland A, Choudhury T (2000) Face recognition for smart environments. Computer 33(2):50–55. https://doi.org/10.1109/2.820039

Chapter 3
Smart Services

Imagine a working scenario…On the envisioned new *Silk Road* we need efficient means of transport on the route between China and Europe. On the first stage the community agents shall recognize the needs and expectations of customers. After completing this task, planning the production and distribution across a network of intelligent agents would then provide for the best solution. In the process *service agents* of the involved customers, producers and consignors are cooperating as fair competitors to fulfill the required tasks. By such a scenario the citizens get the best product, price, time and so forth, according to their demands.

3.1 Introduction

The trends in logistics are transcending from the classical centrally managed material flow from producer to consumer, to a more cellular decentralized supplier-customer structure. In the course of this the logistic system's control functions are being spread all over the network of independent supply chain echelons with their representing autonomous agents. Each echelon's agent shall individually make its own informed decisions, however, they shall be orchestrated to form efficient supply chains, in ways similar to a *beehive* or an *ant colony*.

Web 2.0 shall support such *cellular logistics*. In addition, to provide for the needed transparency as well as time and location accuracy, *smart services* of the *Intelligent Web* shall be used, utilizing advanced *smart sensor networks* as described in the previous chapter. For example, a smart production cell shall communicate and cooperate with other production, distribution and retailer cells, etc. according to the suppliers/customer *peer-to-peer model*, to achieve the common goal—the most flexible and cost-effective service.

R. Gumzej, *Intelligent Logistics Systems for Smart Cities and Communities*,
Lecture Notes in Intelligent Transportation and Infrastructure,
https://doi.org/10.1007/978-3-030-81203-4_3

3.2 Knowledge Web

When talking about the future of the *Internet*, we often speak of the next generation *Web 2.0* or *Intelligent Web* as a *Social Semantic Web* for managing Web content, and the *Gigabyte Internet*, offering unprecedented communication link capacity in terms of availability, communication speed and throughput.

Some also name the new Intelligent Web the *Internet of Things* although it applies to both—social as well as business applications. In fact, the first applications of the Intelligent Web originate from social Web networks (e.g. Facebook, Google+, LinkedIn, etc.). To avoid confusion, a distinction is made between the *Social Web* and the *Semantic Web* which together form the *Intelligent Web*. Hence, the main difference between social and *Semantic Web* from the application point of view is considered the fact that the *Social Web* is mainly exploited by humans, whereas the *Semantic Web* is meant to be exploitable by machines as well.

Both, social and semantic Web share the common concept of *tagging*, however there is much more to the Intelligent Web than tagging itself. It is about building *ontologies* and *metadata* to store and deliver *knowledge* and hereby enable semi-automatic and automatic (autonomous) decision making. The existing *World Wide Web* (*WWW*) is a huge network with connectivity data, metadata, and different links, for different purposes, different users, etc. in which the amount of online information is exponentially growing. To support the goals of the Web 2.0, new intelligent Web solutions are being developed on the basis of *Artificial Intelligence* (AI) and advanced information technologies are being utilized to provide for *data classification*, *data mining* and *intelligent data dissemination*. In the course of this, one speaks about *Web services*, *intelligent agents* and *ubiquitous agent communities*.

According to [1] the *Semantic Web* embodies a vision for a new era in the management and exploitation of Web content by people and machines. According to Net craft's Internet monitoring, the World Wide Web currently comprises more than 190 million Web sites. The rapid growth of information available has not been followed by considerable advances in managing this content. Most of the published data is not structured in a way that would allow for logical reasoning. This makes finding answers that would require more than keyword search difficult.

Tim Berners-Lee considers the *Semantic Web* being about "giving information a well-defined meaning, better enabling computers and people to work in cooperation" [2]. For this purpose he proposed an architecture, based on ontologies and machine-processable metadata [3]. On top of these, the semantic Web contains layers referring to logical reasoning, proof and trust. They exploit the information offered by ontologies and metadata to deliver knowledge and enable automatic and semiautomatic decision-making. The main reasons why these have not been introduced by now are, on one hand the ontologies' top-down modeling approach, which is unnatural to experts from different domains of expertise, and on the other, they are not required for the semantic Web to work. The social Web approach to building the *social semantic Web* seems more natural—letting the users tag and employ their folksonomies for the construction and evolution of ontologies and metadata and hereby deliver semantic Web products and services to users. The main lesson learned here is—simplicity. Simplicity shall prevail over complexity, since simple solutions attract a greater audience, which is needed to get the "big picture" (e.g. *Wikipedia*) complete and organized. This way of "learning", however automated, seems a lengthy process, since the humanity should "reinvent" centuries of its development to achieve this goal, and the driving forces behind this process are weak.

As a conclusion, the new *Web 2.0* shall connect the following *entities*:

- the *humans* in the *social world*,
- *information* and *computing devices* in the *Cyber World*, and
- enabled, *smart devices*, i.e. things that cooperate in the *Physical World* and communicate across the *Cyber World*, constituting the *Internet of Things* (*IoT*).

3.3 Intelligent Agents

According to [4] *agent architectures* are fundamental mechanisms underlying the *autonomous components* that support effective behavior in real-world, dynamic and open environments. In *Artificial Intelligence* (*AI*) an *intelligent agent* is an autonomous entity which observes (through its *sensors*) and acts upon an environment (using its *actuators*; i.e. it is an *agent*) directing its activity towards achieving its goals (i.e. it is *rational*, as defined in economics [5]). Intelligent agents may also learn or use knowledge to achieve their goals. They may be very simple or very complex, depending on the number of their goals and services (sensors and actuators) they require to achieve them.

Russell & Norvig [5] group agents into five classes based on their degree of perceived intelligence and capability:

1. simple reflex agents,
2. model-based reflex agents,
3. goal-based agents,
4. utility-based agents,
5. learning agents.

A *simple reflex agent* acts only on the basis of the current precept from the sensors, ignoring the rest of the precept history (Fig. 3.1). The agent function is based on the *condition-action rule*: "if condition, then action". This agent function only succeeds when the environment is fully observable. In some cases reflex agents can maintain information on their current state, which allows them to disregard conditions, whose actions have already been triggered. Actions are usually in the form of signals to actuators to perform appropriate functions in response to stated conditions.

A *model-based agent* can handle partially observable environments. Its current state is stored inside the agent, maintaining some kind of structure, which describes the part of the system that cannot be perceived directly through its sensors. This

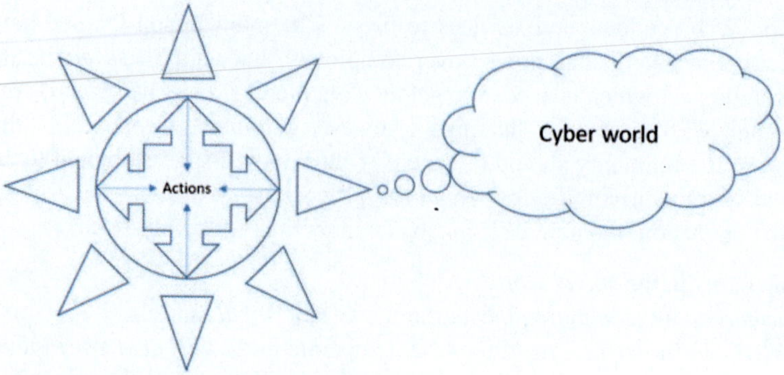

Fig. 3.1 Simple reflex agent

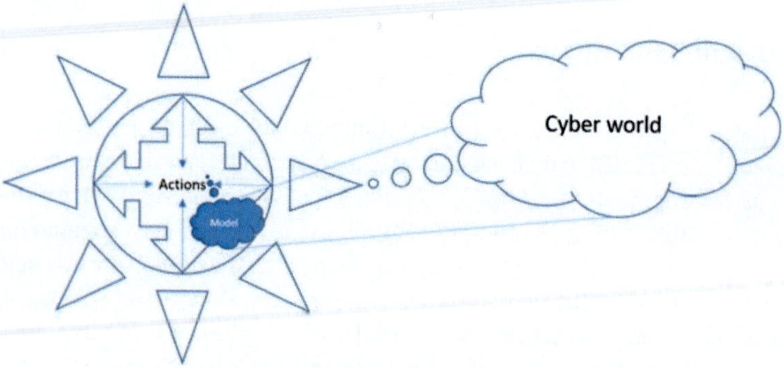

Fig. 3.2 Model-based reflex agent

knowledge about "how the system works" is called the *model of the world*. Hence, such agents are termed model-based agents.

A *model-based reflex agent* should maintain some sort of internal model that depends on the precept history and thereby reflects at least some of the unobserved aspects of the current state (Fig. 3.2). *Precept history* and impact of condition-based actions on the environment can be predicted by using the internal model. Based on these, the agent then chooses the most appropriate action in the same way a reflex agent does. Such an agent may also use models to describe and predict the behaviors of other agents in the environment [6].

A *goal-based agent* further expands the capabilities of a model-based agent, by using a goal information (Fig. 3.3). The *goal information* describes states that are desirable. This allows the agent to choose among multiple possibilities, selecting the one which reaches a goal state. Search and planning are the sub-fields of artificial intelligence destined to finding action sequences that would achieve an agent's goal. Goal-based agents only distinguish between goal states and non-goal states.

A *utility-based agent* in addition to defining goal states also defines a measure of how favorable a particular goal state is. This measure can be obtained through the

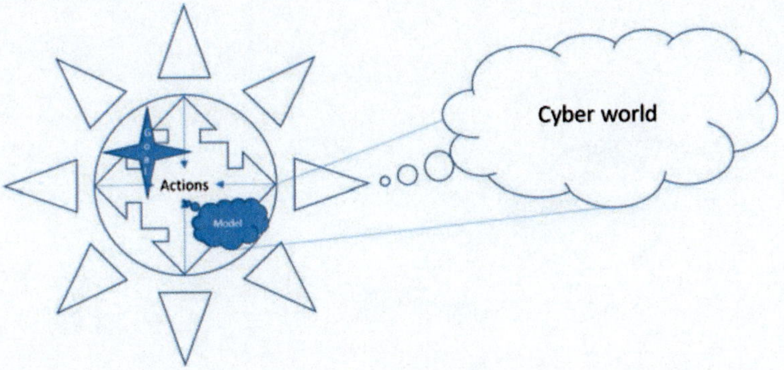

Fig. 3.3 Goal-based agent

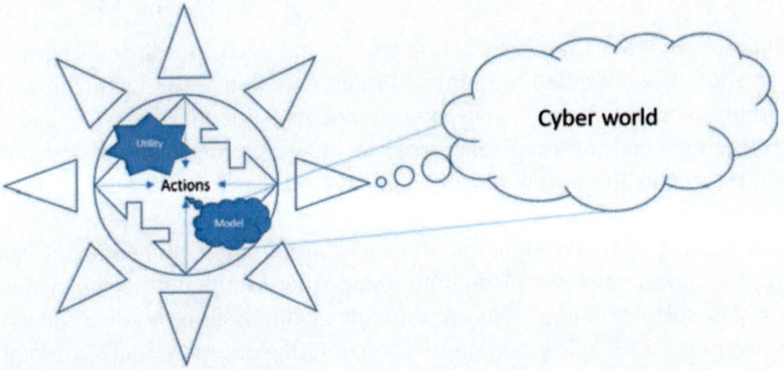

Fig. 3.4 Utility-based rational agent

use of a *utility function* which maps a state to a measure of the utility of the state. A more general performance measure should allow a comparison of different system states according to their preference when achieving an agent's goals. The term *utility* can be used to describe its success in pursuing its goals.

A *utility-based rational agent* chooses the action that maximizes the expected utility of the action outcomes—that is, what the agent expects to achieve, on average, given the condition probabilities and utilities of each outcome (Fig. 3.4). A rational utility-based agent needs to model and keep track of its environment, as well as tasks that have involved a great deal of research on perception, representation, reasoning, and learning.

With *learning agents* their learning property has the advantage that it allows such agents to initially operate in unknown environments and then gradually become more competent, as their knowledge grows (Fig. 3.5). Here, the most important distinction is made between the *learning element*, which is responsible for making improvements in its model of the world, and the *performance element*, which is responsible for selecting actions based on their anticipated effect on the external world. The learning

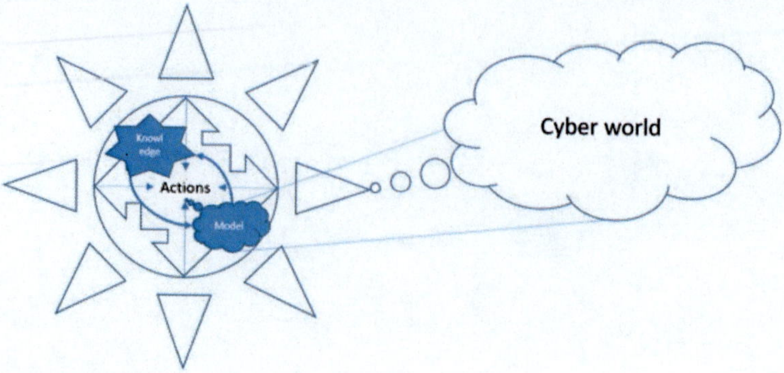

Fig. 3.5 Learning agent

element uses feedback from agent action's success to determine how the performance element should be modified to improve its future actions. The main component of the learning agent is a *problem generator*. It is responsible for suggesting actions that will lead to new and informative experiences. On the other hand, the performance element represents the action part of an agent—it takes in precepts and decides on actions.

Agents can be organized into hierarchies. A *multi-agent system* (*MAS*) is a computerized system composed of multiple interacting intelligent agents. Multi-agent systems can solve problems that are difficult or impossible to solve for any individual agent or a monolithic system. Their intelligence may include methodical, functional, procedural approaches, algorithmic search or reinforcement learning.

Intelligent agents are often applied as automated online assistants. They perceive the needs of customers in order to perform individualized customer services. Such agents typically consist of a dialog system, an avatar, as well an expert system to provide for specific expertise to their users. They can also be used to optimize the coordination of their users in the (cyber) world.

As mentioned earlier, agents in the cyber world not only represent humans but can also represent smart devices in the IoT. Hence, we may conclude that entities in the cyber world may employ agents to cooperate and fulfill their goals. By doing so they utilize their services and the agents can utilize the services of other agents of a MAS to perform their actions. According to the *Foundation for Intelligent, Physical Agents* (*FIPA*), each agent has its own characterization, which represents its behavior. While intelligent agents only operate with a limited degree of independence, *autonomous agents* on the other hand may automatically act on owner's behalf and carry out their plans without any user interference. Hence, they represent a *self-organizing system*.

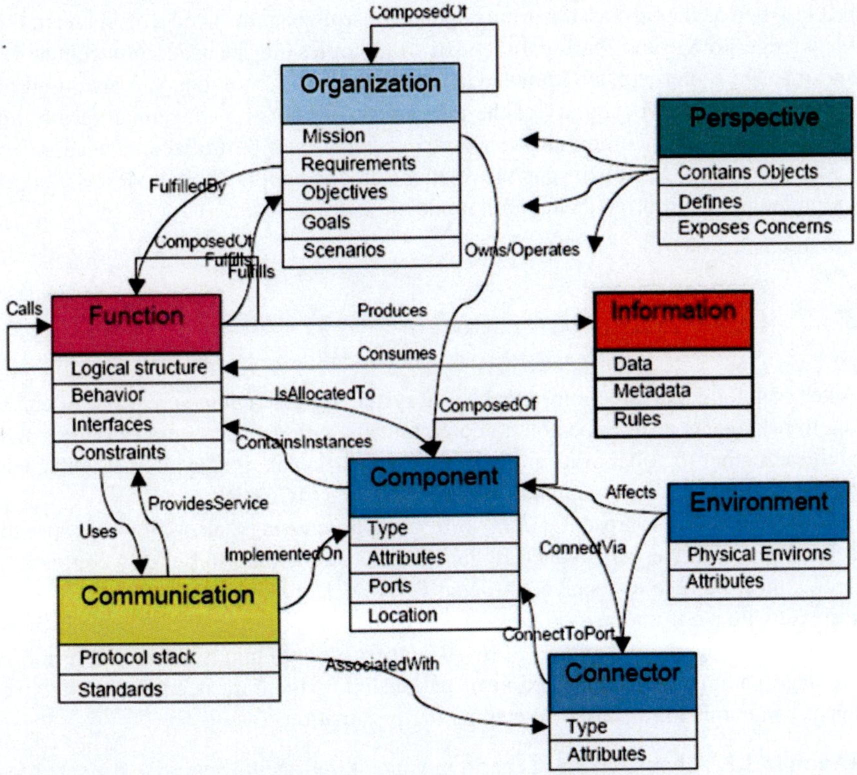

Fig. 3.6 Top level ontology based on the nominal set of views; picture by Peter Shames & Joseph Skipper. NASA, JPL., Public domain

3.3.1 Knowledge Sharing

Each multi-agent system needs a mechanism for *knowledge sharing*. According to [7] *ontologies* were developed in *Artificial Intelligence* (*AI*) to facilitate knowledge sharing and reuse. They enable a shared and common understanding of a defined domain that can be communicated between people and application systems. *Agent communication protocols* may be utilized to access this knowledge and to establish sensible relations among entities in the cyber world as well as to implement appropriate behavioral models. In Fig. 3.6 the top level ontology of a Framework for Modeling Space Systems Architectures [8] is outlined as an example of an all encopassing base ontology for open and shared environments.

To apply the intelligent MAS decentralized agent-based approach to e-marketplace automation for example (Chap. 17), any pair of supply chain partner's agents should be autonomously integrated into both—the information and material flows—with the goal of meeting supply chain nodes' goals. By their behavior, such agents would be classified as model-based learning agents. Learning agents need a mechanism for knowledge storage and sharing. In logistics, ontologies may be used to conceptualize and manage e-marketplace knowledge (cp. Fig. 17.3). Ontologies are also used to establish the basic vocabulary of the relevant supply chain management terms and transactions. Agent communication protocols may then be utilized to access this knowledge and to establish sensible relations among supply chain nodes as well as to implement appropriate behavioral models.

3.3.2 Communication

A key component of any multi-agent based system is *agent communication*. According to [4], agents need to be able to communicate with users, system resources and with each other to collaborate and negotiate. In particular, agents interact with each other by using the *Agent Communication Language* (*ACL*) [9].

Agent messages represent *actions* or *communicative acts*, also known as speech acts or performatives. It is stated in the FIPA standards that, to be fully compliant, agents must be able to receive any legal FIPA-ACL message and at the very least respond with a not-understood message, in case the processing of the message does not render a meaningful result. FIPA-ACL messages contain a set of one or more message parameters. The selection of parameters, which are needed for effective agent communication, varies depending on the situation.

Example 3.1 A book buying scenario in which three agents negotiate the sale (cp. Fig. 3.7). First a yellow-page service is contacted by the customer agent (c) to determine whether there are providers available. In response to this request, the yellow-page service notifies the customer agent that there are two sellers of the requested item (s1 and s2). In the sequel the transactions are performed according to a common requisition scenario in supply chain management of sending a request for quotations and deciding on the best offer to make the deal:

Example 3.2 Book trading scenario ACL messages (cp. Fig. 3.7):

```
(REQUEST  :sender  (agent-identifier :name c@192
    .168.1.101:1099/JADE)  :receiver  (set (agent-
    identifier :name df@192.168.1.101:1099/JADE))  :
    content  ``((action (agent-identifier :name df@192
    .168.1.101:1099/JADE) (search (df-agent-description :
    services (set (service-description :type item-selling)
    )) (search-constraints :max-results -1))))''  :reply-
    with  rw-c@192.168.1.101:1099/JADE1620278614707-2  :
    language  fipa-sl0  :ontology  FIPA-Agent-Management
    :protocol  fipa-request  :conversation-id  conv-c@192
    .168.1.101:1099/JADE1620278614707-2)

(INFORM  :sender  (agent-identifier :name df@192
    .168.1.101:1099/JADE)  :receiver  (set (agent-
```

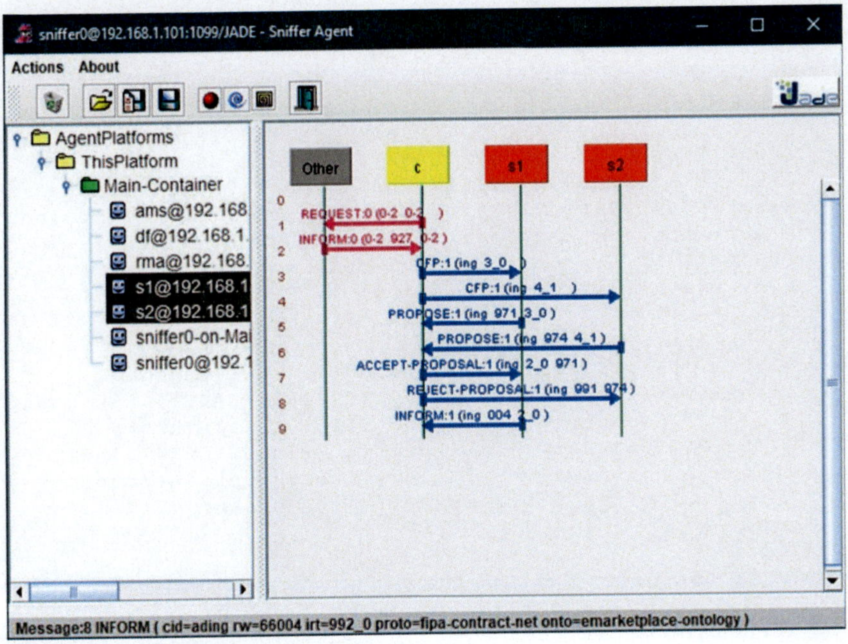

Fig. 3.7 Book trading scenario in JADE framework

```
identifier :name c@192.168.1.101:1099/JADE)) :content
  ''((result (action (agent-identifier :name df@192
.168.1.101:1099/JADE) (search (df-agent-description :
services (set (service-description :type item-selling)
)) (search-constraints :max-results -1))) (sequence (
df-agent-description :name (agent-identifier :name
s2@192.168.1.101:1099/JADE) :services (set (service-
description :name item-trading :type item-selling))) (
df-agent-description :name (agent-identifier :name
s1@192.168.1.101:1099/JADE) :services (set (service-
description :name item-trading :type item-selling)))))
)''  :reply-with  c@192.168.1.101:1099/
JADE1620278614716  :in-reply-to  rw-c@192
.168.1.101:1099/JADE1620278614707-2  :language  fipa-
sl0  :ontology  FIPA-Agent-Management  :protocol  fipa
-request  :conversation-id  conv-c@192.168.1.101:1099/
JADE1620278614707-2)

(CFP  :sender  (agent-identifier :name c@192
 .168.1.101:1099/JADE)  :receiver  (set (agent-
 identifier :name s2@192.168.1.101:1099/JADE))  :
 content  ''((INQUIRY (ITEM :name book) 2))''  :reply-
 with  R1620278614753_0  :language  fipa-sl0  :ontology
   emarketplace-ontology  :reply-by  20210506
```

```
    T052243756Z  :protocol  fipa-contract-net  :
    conversation-id  item-trading)

(CFP  :sender  (agent-identifier :name c@192
    .168.1.101:1099/JADE)  :receiver  (set (agent-
    identifier :name s1@192.168.1.101:1099/JADE))  :
    content  "((INQUIRY (ITEM :name book) 2))"  :reply-
    with  R1620278614757_1  :language  fipa-sl0  :ontology
      emarketplace-ontology  :reply-by  20210506
    T052243756Z  :protocol  fipa-contract-net  :
    conversation-id  item-trading)

(PROPOSE  :sender  (agent-identifier :name s2@192
    .168.1.101:1099/JADE)  :receiver  (set ( agent-
    identifier :name c@192.168.1.101:1099/JADE))  :content
      ''5''  :reply-with  c@192.168.1.101:1099/
    JADE1620278614766  :in-reply-to  R1620278614753_0  :
    language  fipa-sl0  :ontology  emarketplace-ontology
    :protocol  fipa-contract-net  :conversation-id  item-
    trading)

(PROPOSE  :sender  (agent-identifier :name s1@192
    .168.1.101:1099/JADE)  :receiver  (set (agent-
    identifier :name c@192.168.1.101:1099/JADE))  :content
      ''10''  :reply-with  c@192.168.1.101:1099/
    JADE1620278614771  :in-reply-to  R1620278614757_1  :
    language  fipa-sl0  :ontology  emarketplace-ontology
    :protocol  fipa-contract-net  :conversation-id  item-
    trading)

(ACCEPT-PROPOSAL  :sender  (agent-identifier :name c@192
    .168.1.101:1099/JADE)  :receiver  (set (agent-
    identifier :name s2@192.168.1.101:1099/JADE))  :reply-
    with  R1620278614786_0  :in-reply-to  c@192
    .168.1.101:1099/JADE1620278614766  :language  fipa-sl0
      :ontology  emarketplace-ontology  :protocol  fipa-
    contract-net  :conversation-id  item-trading)

(REJECT-PROPOSAL  :sender  (agent-identifier :name c@192
    .168.1.101:1099/JADE)  :receiver  (set (agent-
    identifier :name s1@192.168.1.101:1099/JADE))  :reply-
    with  s1@192.168.1.101:1099/JADE1620278614785  :in-
    reply-to  c@192.168.1.101:1099/JADE1620278614771  :
    language  fipa-sl0  :ontology  emarketplace-ontology
    :protocol  fipa-contract-net  :conversation-id  item-
    trading)

(INFORM  :sender  (agent-identifier :name s2@192
    .168.1.101:1099/JADE)  :receiver  (set (agent-
    identifier :name c@192.168.1.101:1099/JADE))  :reply-
    with  c@192.168.1.101:1099/JADE1620278614798  :in-
    reply-to  R1620278614786_0  :language  fipa-sl0  :
```

```
ontology emarketplace-ontology :protocol fipa-
contract-net :conversation-id item-trading)
```

In the above example (cp. Fig. 3.7), autonomous agents communicate their messages over the HTTP(S) protocol based on the FIPA-contract-net Protocol [9]. It allows one agent (*Initiator*), to have some tasks performed by one or more other agents (*Participants*) as well as further optimization of the functions that characterize the task. By initiating a procurement action, a number of eligible participants, paged with the *Call for Proposals* (*CFP*) message, may respond with a proposal, while the rest must refuse. The Initiator would then continue negotiations with the participants that proposed. Finally, the selected supplier agent would be notified and fulfill the order by informing the customer agent that the order has been confirmed.

3.3.3 Behavior

All agents are generic until they are associated with an ontology. In case of e-marketplace automation (Chap. 17) they are assigned their role and behavior by the e-marketplace ontology (cp. Fig. 17.5). Initially an agent is registered with a yellow page service, considering its specific role (customer, supplier, customer & supplier). After that, every pair of nodes that interact with each other within the e-marketplace can be observed in terms of a customer-supplier relation. After being triggered by a Kanban signal, for example, a customer agent may automatically order an item (material, service or finished product) from its supplier(s) agents based on its availability, price and other criteria.

Let us consider a typical ordering operational cycle from the perspective of agent behavior (cp. Fig. 17.6):

1. The customer posts a call for proposals (CFP) message for the item to all known suppliers; it contains the requested number of items and date of order fulfillment.
2. Suppliers either PROPOSE by their offers and provide the data on the price for the requested amount of items or REFUSE the request.
3. The customer then makes the decision on the acceptance of a particular offer, based on the prices and some additional service quality criteria from all suppliers that provided positive feedback to its inquiry.
4. An order is then made to the selected supplier by posting it an ACCEPT-PROPOSAL message, while a REJECT-PROPOSAL message is sent to all other participating suppliers.
5. The order is then either acknowledged as being fulfilled or a failure to fulfill the order.
6. Finally, service quality indicators of the participating supplier agents are updated, based on the outcome, to be considered in the next ordering operation.

The described behavior of supply chain agents augments the FIPA-contract-net Protocol. In addition their learning and performance elements are depicted by the knowledge base of prospective supply chain partners and their quality of service. Since any supply chain agent strives to improve its quality of service, it represents the utility function on its performance for achieving its goals (cp. Chap. 18).

3.4 Autonomous Systems

Let us extend our consideration of autonomous agents to autonomous systems. An autonomous system (*AS*) is a *Cyber Physical System* represented by at least one *autonomous agent*. Depending on the number of its *roles* and *overlay networks* of coordinated *autonomous systems*, where it is a member, it may employ a corresponding number of agents.

3.4.1 Overlay Networks

Overlay networks (Fig. 3.8) are dedicated networks of *autonomous systems* in the *cyber physical space*, representing physical application specific networks. They may be relatively simple (e.g. intelligent sensor networks) or quite complex (e.g. e-marketplace networks, smart traffic networks, e-health networks, electrical distribution networks, etc.).

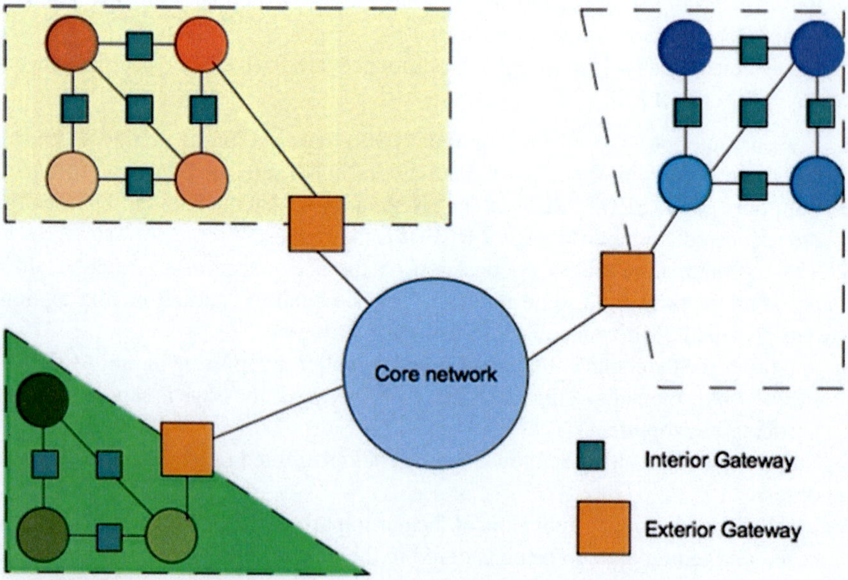

Fig. 3.8 Overlay network of cooperating autonomous systems

The properties of overlay networks and the properties of the participating autonomous system's agents are defined by their *ontologies*. Since they tend to be quite sparse, they usually involve local *interior gateways* acting as primary interrogators for the actions performed by local agents, and *exterior gateways* as secondary interrogators to a wider, i.e. regional or global *cyber space* of agents (cp. Fig. 3.8) of the same network. Hereby, overlay networks of coordinated autonomous systems are formed in the cyber space to accurately represent complex physical systems like a *Smart City* (cp. Chap. 20) or complex business networks like a *supply network* (cp. Chap. 17).

Autonomous systems nodes constitute a Peer-to-Peer (P2P) network (Fig. 3.10). Their knowledge base (Fig. 3.11) supports local decision-making, as inspired by the Kleinberg's Small World Model [10], with the goal of fulfilling goals and ensuring sustainable configuration management throughout the entire overlay network.

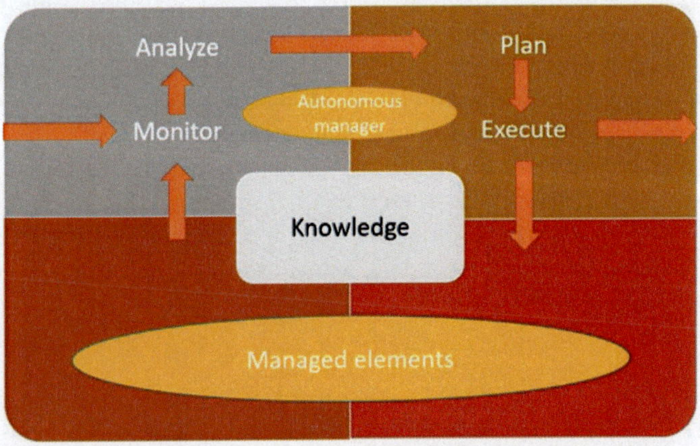

Fig. 3.9 Autonomous elements consist of manager and managed components

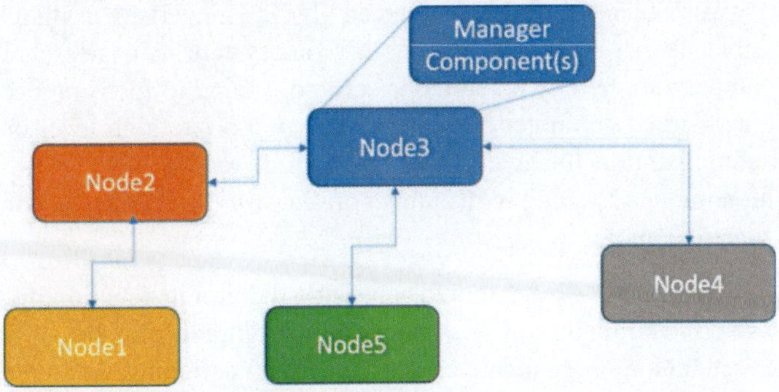

Fig. 3.10 Autonomous system's nodes network

3.4.2 Composition

Every *autonomous system* (*AS*) (Fig. 3.10) consists of autonomous elements (Fig. 3.9) like any *overlay network* (e.g. traffic, supply, power distribution network, etc.) is composed of nodes (Fig. 3.10).

Every AS node consists of at least one *managed element*, being controlled and represented by their *autonomous manager*. A *managed element* typically represents an existing (software) component of a computing system (e.g. database server, web service, ERP system) that can be made autonomous.

The main functions of an *autonomous manager* are *monitoring, analysis, planning and executing* [11]. As it is the job of a supply chain manager, for example, to make informed decisions on when and how to produce as well as when and where to order to fulfill expected demand based on inputs from the underlying ERP system and offers from the SCM system, it is the task of the autonomous manager to monitor inputs from the system and its environment, analyze them and devise a feasible execution plan that influences both its managed elements as well as associated nodes. The autonomous manager operation cycle correlates with the *MAPE-K control loop*, which is a typical working scenario of an *autonomous agent*.

Eventually, the border between manager and managed components of an AS node is expected to disappear, rendering fully integrated well defined AS nodes (behavior, interfaces) with few restrictions to their internal structure. The concepts of *GRID computing* and *Web services* play a key role in their development where the individual elements would take over the function of autonomous agents that implement certain services by their collaboration within their network.

3.4.3 Properties

The *autonomous systems paradigm* arose as a result from the "computer software complexity crisis" that was first "announced" by P. Horn, IBM in march 2001 at Harward university: "As our autonomous nervous system manages our heart beat and body temperature hereby leaving our conscious mind to focus on our tasks, so should the networked computer systems, given appropriate high level instructions from their administrators, be able to manage themselves".

Thus autonomous systems have a number of *self-properties* that characterize them as such:

Self−management : An AS should constantly monitor its functioning, check for updates and install them, roll back previous configurations in case it determines that the installation of an update caused unstable functioning, ...

Self−configuration : AS will configure themselves automatically in correspondence with high-level business rules, prescribing the desired goal and without precise instructions on how it should be achieved. A new AS would seamlessly integrate itself into its environment that shall adapt to its presence as well—by

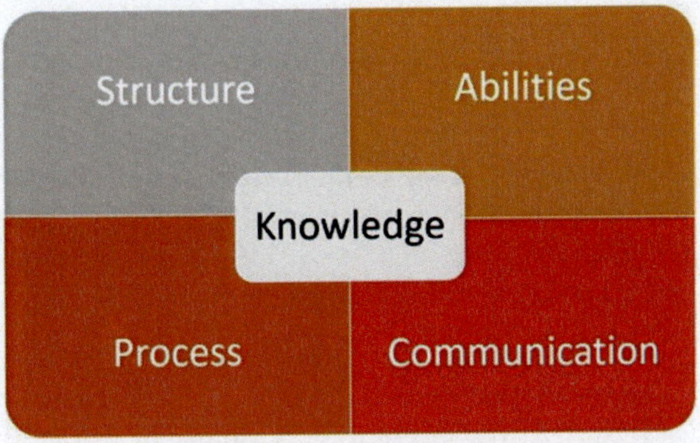

Fig. 3.11 Autonomous system's nodes capabilities

registering its capabilities (Fig. 3.11), enabling it to be recognized as such by its environment.

Self-optimization: AS will constantly seek ways to improve their efficiency.

Self-healing: AS will diagnose flaws themselves and fix or isolate (for maintenance) their ill-performing elements.

Self-protection: AS shall self-protect themselves in two ways: (1) by self-healing to prevent system failures and (2) by early detection of potential risks and launching appropriate protective measures to prevent attacks.

Although the *AS* would eventually be able to overcome the majority of systems' integration and maintenance efforts, the humans shall still need to provide for the high level *rules—goals* and *limitations* to their functioning. For example, the business logic of supply chain, power distribution or traffic management would have to be described formally to autonomously operate these overlay networks. These rules need to be correct and unambiguous, since the consequences of erroneous or ambiguous rules are unpredictable and hard to repair. The key challenge to defining the rules is to ensure their coherency, consistency, and safety, even though they might not be entirely defined (a human property). It is not mandatory to solve the problems of artificial intelligence (AI) to implement AS, although previous considerations point to it.

First examples of AS have already been introduced in the form of *smart devices* (e.g. smart phones, smart glasses, smart homes, etc.) and *Web services* with their *Cloud applications* and *mobile apps*.

3.5 Synopsis

In relation with the aforementioned arising smart technologies, as always, new opportunities and issues alike shall arise as well. The envisaged applications of *extended reality* in the future *Innovation Society* are considered a prominent example of the use of *smart devices* on one and *smart services* on the other side (cp. Chap. 13).

Data safety and security is a pressing issue facing governments, businesses and individuals. Enormous amounts of data are generated and transmitted daily and their users must be confident that unwanted third parties in no way have access to their transactions. Greater demands are being placed on technology as we move to smaller, wireless devices using as little power as possible. In addition, major industrial and infrastructural systems are increasingly being controlled via the Internet. Such systems can be vulnerable to outside attacks. To provide for safety and security of devices and services in the arising *Internet of Things*, two main questions need to be answered:

1. How can we ensure the accuracy and consistency of their data?
2. How can we constrain the visibility of private information to their users and stakeholders?

From the safety point of view, the obvious answer is—by correctly managing and replicating them. From the security point of view, the answer at hand is—by authentication and authorization. But, how can we ensure proper data replication, authentication/authorization of users and devices and avoid stolen identities? The next chapter elaborately deals with these and related questions.

References

1. Mikroyannidis A (2007) Toward a social semantic web. Computer 40(11):113–115
2. Berners-Lee T, Hendler J, Lassila O The semantic web. https://www.scientificamerican.com/article/the-semantic-web/
3. Berners-Lee T (2000) Semantic web—xml2000. https://www.w3.org/2000/Talks/1206-xml2k-tbl/
4. Bellifemine FL, Caire G, Greenwood D (2007) Developing multi-agent systems with JADE. (Wiley Series in Agent Technology). Wiley, Hoboken, NJ, USA
5. Russell S, Norvig P (2009) Artificial intelligence: a modern approach, 3rd edn. Prentice Hall Press, USA
6. Albrecht SV, Stone P (2018) Autonomous agents modelling other agents: a comprehensive survey and open problems. Artif Intell 258:66–95
7. Fensel D (2011) Ontologies, pp. 11–18. Springer Berlin Heidelberg, Berlin, Heidelberg. https://doi.org/10.1007/978-3-662-04396-7_2
8. Shames P, Skipper J (2006). Toward a framework for modeling space systems architectures. https://doi.org/10.2514/6.2006-5581
9. Foundation for Intelligent Physical Agents: Fipa 97 specification—part 2—agent communication language. http://www.fipa.org/specs/fipa00018/OC00018.pdf
10. Kleinberg J (2000) The small world phenomenon: an algorithnic perspective. In: Proceedings of the 32nd ACM symposium theory of computing, pp 163–170
11. Gumzej R, Sukjit P, Unger H (2012) Modeling overlay networks for autonomous supply chain systems. Logist Sustain Transp, Celje, Slovenia 3:41–50

Chapter 4
Safety and Security

In the "digital era" two notions have become critical—*information safety and security*. On one hand, having the correct data in the correct format when one needs them is more important and easier to achieve than ever before by replicating the data, storing it in a framework of a reliable cloud service (e.g. Amazon, Dropbox, Google Drive, etc.).

On the other hand, even when transmitting the data to a reliable storage space, one should insist on data transmission through secure channels in order not to be intercepted, viewed or even modified by third parties. Furthermore, one should also insist on keeping the private data private, since not all data should be accessible to everyone. Not even all authorities may have the same access rights. Besides personal data, sensitive information, only required to run a public service, production plant, or public transport network are good examples thereof.

Of course, in order to achieve this, employing sound *safety and security policies* in order to separate the private from the public data and services and appropriate *authentication* and *authorization* of users and their agents, are a must.

4.1 Introduction

According to [1], computer-run control systems nowadays do everything from regulating temperature, electrical current, pressure and flow rate to ensure the proper functioning of dams, factories, nuclear power plants and other essential parts of a country's economy and infrastructure. These systems are both more convenient and less labor-intensive, but also more vulnerable to hackers and terrorist attacks.

The two major types of these systems are *Supervisory Control And Data Acquisition* (*SCADA*) systems and *Process Control Systems* (*PCS*s). SCADA systems are more commonly found in utility and other infrastructure operations, while PCSs are

R. Gumzej, *Intelligent Logistics Systems for Smart Cities and Communities*,
Lecture Notes in Intelligent Transportation and Infrastructure,
https://doi.org/10.1007/978-3-030-81203-4_4

more often found in production facilities. SCADA systems typically track data and sound alarms when a hazardous conditions arise. PCSs, on the other hand, handle more complex information, make decisions and take action, as situations require. Neither, however, commonly have built-in safety or security features, unless these requirements were an original part of their design specifications. Instead, they mostly rely on replicated and fallback systems to provide for safety and passwords or physical keys for security. In addition, the devices, which make up such systems, are often made up of a large variety of components from different producers, which were not designed to work together, and thus provide no guarantees for either safety or security. A common hacker attack aimed at these types of systems is the *Denial-of-Service* (*DoS*) attack. In a DoS attack, a hacker floods a system with useless information, which overwhelms the system's sensors, making it unable to perform its primary functions.

Several high-profile hacker attacks in the last two decades have brought these concerns to the forefront. For example, in 2000 a hacker attacked the SCADA system of an Australian sewage treatment facility, causing a full blockage of sewage in a city in Queensland, using a remote radio transmitter. In January 2003 the Slammer worm was used to attack the safety monitoring system of the US Davis-Besse nuclear power plant, by accessing the plant's network through a telephone, which was connected to the infected computer of a contractor. Many attacks are never made public because the victims do not want the general public to know that their systems were compromised and may hence be considered vulnerable. On the other hand, increasing security is a double-edged sword. If securing (or "hardening") makes them more cumbersome or hard to access to their daily users, it increases the likelihood that they will find shortcuts or workarounds across the security mechanisms, leaving the system even more vulnerable. Nowadays, more sophisticated methods of attacks are used for targeting public audience, which ensure criminals to get what they want and get away with it undetected.

Devising new generations of *data security solutions* is an on-going challenge in the information age. In this chapter some of the most promising methods and mechanisms to be used in the arising *Information Society 2.0* and *Industry 4.0* are presented. In the future green smart cities, various combinations of the security solutions presented here will play a crucial role in all aspects of establishing safe and secure application frameworks for personal and professional use. In the sequel the mentioned technologies and their security features are presented in more detail.

4.2 Communication Networks Security

4.2.1 WiFi Networks

The increasing popularity of WiFi networks based on IEEE 802.11b/g WLAN standards gives hackers extraordinary opportunities to access valuable, sensitive personal and corporate information. Within the corporate premises, the wireless access points of a company can be monitored and protected by professionals from a company's IT department. What happens, though, when an employee needs to access the company network from a remote location through a home or even public WiFi network? Corporations spend great deals of money ensuring the security of their networks, but can they expect their employees to do the same to protect their home networks? Can public WiFi networks even obtain the same level of protection? The obvious answer is no.

Wired Equivalent Privacy (*WEP*) is the encryption mechanism originally designed for WiFi security, but it has already been cracked by hackers [2]. If WEP is the only encryption system used on a WiFi network, hackers have readily available tools to access not only the wireless network, but also the wired network behind the wireless access points. This allows hackers to obtain all sorts of information from private or corporate networked computers and enables them to upload anything from Spam to Trojan horses [2]. Meanwhile, there are better alternatives to WEP such as the *WiFi Protected Access*4 (*WPA*), *Virtual Private Networks* (*VPN*) and *captive portals*. WPA and VPN have some configuration and interoperability problems, but do show potential to become the ultimate network security solutions. Captive portals, often used by hotels with *pay-for-use WiFi* networks, can only be assessed on a case-by-case basis because each portal's cryptographic and security protocols must be examined. Until a comprehensive solution to protect from unauthorized wireless network access is available, individuals and corporations must carefully consider the ramifications of WiFi use.

4.2.2 RFID and Ad-hoc Networks

Wireless Sensor Networks (WSN) and *Radio Frequency Identification* (RFID) devices are extremely small, low power computing devices, which run applications, used in diverse industries from healthcare to logistics. In all mentioned cases, privacy and trust are imposed by the use of *cryptographic algorithms*, designed to ensure *data confidentiality*. Concerning network topologies in Wireless Sensor Networks, two main concepts compete: *ad-hoc networks* (IEEE 802.11p) and *cellular networks* (e.g. *cellular vehicle to everything*, *C-V2X*). Both concepts imply technical advantages and disadvantages and are subject to commercial and legal constraints.

A major challenge for engineers is how to provide for complex, ultra-secure encryption algorithms, which are also lightweight—requiring a minimum amount of energy. Today's wireless sensor nodes are mostly battery-powered. However, engineers are already developing new generations of sensor nodes, which will be powered by harvesting energy from their surroundings, possibly storing it for later use, rather rely solely on their batteries. Such self-powered devices are known as *energy scavengers* and can power themselves by converting heat, light and vibrations into electricity. Contemporary energy scavengers are able to produce only a fraction of the power required to support today's CPUs running in sensor networks [3].

Though the computational power requirements of wireless sensor nodes are small, compared to those of transmission power, *cryptography overheads* must still be minimized. In order to do this, algorithms must be designed for *ultra low-power applications*. Popular algorithms, currently being used, include *block ciphers, stream ciphers, hash functions* and *public-key cryptosystems*. None of these are without drawbacks. Thus, it is necessary to develop new algorithms which would enable simple and efficient authentication and authorization protocols to be used in ultra low-power applications.

4.3 Personal Security

To ensure personal safety and security, according to [4], access to reliable personal identification infrastructure is becoming essential. Conventional methods of identification based on the possession of ID cards or exclusive knowledge—like a social security number or a password—are not altogether reliable. ID cards can be lost, forgot, or misplaced; passwords can be forgotten or compromised.

Biometric technology is becoming a viable alternative to traditional identification systems in many government and commercial application domains. According to [5], one can conclude that biometric passwords will soon replace their alphanumeric counterparts with *biometric signatures* which cannot be stolen, forgotten, lost or lent to another person. A *biometric system* is a pattern recognition system that establishes the authenticity of specific physiological or behavioral characteristics possessed by a user [4].

Logically, any biometric system can be subdivided into two parts or stages:

1. the enrollment module/mode, responsible for training the system to identify a given person, and
2. the identification module/mode, responsible for recognizing the person.

4.3.1 Biometrics

Biometrics are technologies that automatically confirm the identity of people by comparing the patterns of physical or behavioral characteristics in real time against enrolled computer records of those patterns. Contemporary biometric technologies accomplish this task by scanning in patterns of a person's face, fingerprint, hand, iris, palm, signature, skin, or voice. Biometrics uses methods for unique recognition of humans based on one or more intrinsic physical or behavioral traits. A person's *physiological or behavioral characteristics*, known as *biometrics*, are crucial for his/her *identification* and *authentication*. The biometric approach to authentication is appealing because of its convenience and possibility to offer strong security with non-repudiation. However, specialized hardware such as *biometric scanners* and complex software for *feature extraction* and *biometric template matching* are required.

4.3.2 Biometric Security Mechanisms

There are two kinds of biometric systems: authentication and authorization. In *authentication systems* a biometric signature of an unknown person is presented to a system. The system compares the new biometric signature with a database of biometric signatures of known individuals. On the basis of the comparison, the system checks the identity of the unknown person against its database. Systems that rely on authentication include those that the police uses to identify people from fingerprints and mug shots [6].

In *authorization systems*, a user presents a biometric signature with a claim that a particular identity belongs to the concrete biometric signature. Authorization applications include those that verify identities during point-of-sale transactions, access control to restricted areas, such as warehouses, labs, access to health records, bank accounts, etc.

Among the many authentication systems that have been proposed and implemented, *finger vein biometrics* is emerging as the foolproof method of automated personal identification. Finger vein is a unique physiological biometric for identifying individuals based on the physical characteristics and attributes of the vein patterns on a human finger. It is a fairly recent technological advance in the field of Biometrics that is being applied to different fields such as medical, financial, law enforcement facilities and other applications where high levels of security or privacy are very important [7]. A fingerprint is the pattern of ridges and valleys on the surface of a fingertip. Each individual has unique fingerprints. Most fingerprint matching systems are based on four types of fingerprint representation schemes (cp. Fig. 4.1): gray-scale image [8], phase image, skeleton image [9], and minutiae [10, 11]. Due to its distinctiveness, compactness, and compatibility with features used by human fingerprint experts, the minutiae-based representation has become the most widely adopted fingerprint representation scheme.

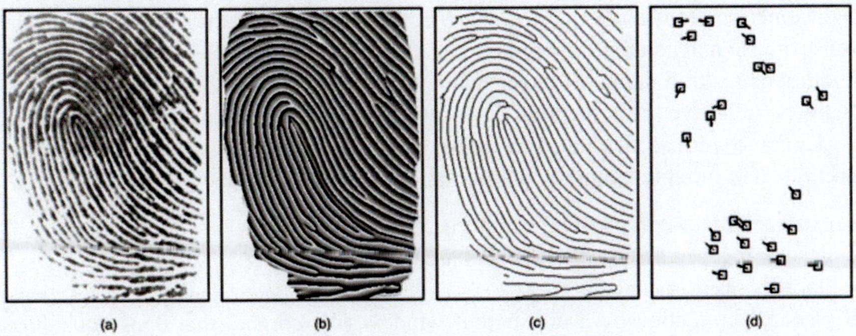

Fig. 4.1 Fingerprint representation schemes: **a** gray-scale image (FVC2002 DB1, 19_1), **b** phase image, **c** skeleton image, and **d** minutiae [12]

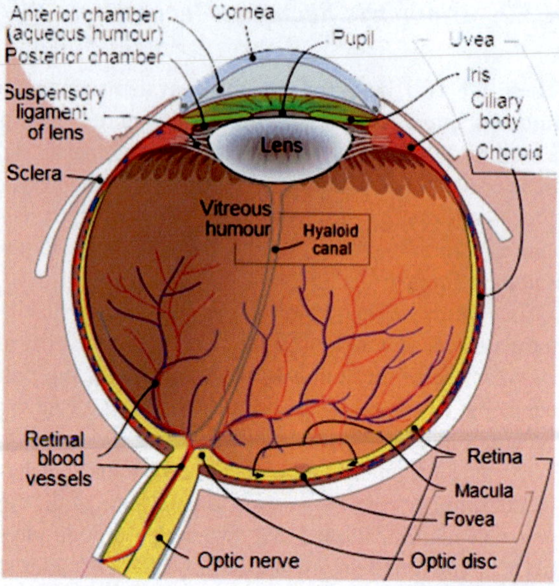

Fig. 4.2 A schematic picture of the human eye; picture by Rhcastilhos. And Jmarchn., distributed under CC https://creativecommons.org/licenses/by-sa/3.0 BY-SA 3.0 license

Other, even more secure but also more cumbersome methods of biometric authentication are *retinal and iris identification* (cp. Fig. 4.2). Retinal blood vessel pattern is unique for each human being, even in the case of identical twins. Moreover, it is a highly stable pattern over time and totally independent of genetic factors [13]. Also, it is one of the hardest Biometrics to forge as the identification relies on the blood circulation along the vessels. These properties make retinal identification one of the best biometric authorization methods for high security environments.

Similar to retinal identification, *iris biometric identification* requires a special eye-scanner. The iris is a small connective tissue and muscular structure of around 12mm diameter with a central opening called the pupil. It controls the amount of light, entering the eye, which is focused by the lens onto the retina so as to provide the sense of vision. It contracts in bright light making the pupil smaller and dilates in dark conditions making the pupil larger, which together with the source of the incident light can influence the perception of an individual's eye color and iris pattern [14].

Similar to retinal identification, an iris scan may be considered an exceptionally accurate and suitable biometric authorization method due to its:

- extremely data-rich physical structure,
- genetic independence—no two eyes are the same,
- stability over time, and
- physical protection by a transparent window (the cornea) that does not inhibit external view-ability.

4.4 Objects, Vehicles and Facilities Security

The automatic identification procedure for objects and facilities, which has been dominating the world for the last 40 years, namely bar-code, has proven useful for most purposes, however it has some deficiencies. The most important two of them are its need for a clean line of sight and the fact that it can be easily tampered—made unreadable or misinterpretable—by simple wear-out, covering with paint or by double labeling. The main reason, why bar-codes are so popular, is the fact that the barcode labels are low cost and bar-code scanners are wide spread. Hence, sustaining a one-way tracking system is relatively easy, provided a label can be applied to an object. Due to wear-out reverse tracing is relatively difficult, since the labels may drop off or become unreadable even due to a thin film of dust. Due to the increasing need for green—inverse—logistics, an identification technology that could outlive the labeled objects is needed. Depending on the nature of the data and its carriers the security of their identity and data might be critical. Latest developments in the field of *Data Matrix* technology [15] provide for improved data integrity and enable data encryption, leaving the rest of the optical identification problems unresolved.

The answer to most concerns in relation to barcode technology is *Radio Frequency Identification* (RFID). For this reason it is also considered a successor to barcode, although both identification technoques are still evolving individually. Initially, RFID also had problems, since electromagnetic waves, used in automated identification, are also prone to errors and no battery would last for more than a few years. However, one came up with multiple solutions, the simplest and cheapest of which could easily substitute any barcode.

An *RFID label* (passive tag) doesn't need its own power supply, because it uses contactless energy from the reader [16]. It has all the mentioned benefits over the bar code with just a slightly bigger price tag, which is also getting lower. Another important benefit is the fact that the use of RFID technology is practically unlimited—one can use it almost everywhere—in logistics, military, healthcare, banking, etc. [17] In the modern global Information Society there is an ultimate concern for privacy that enables only authorized readers to identify and read the received data. Hence, this might also be the ultimate quest of RFID technology to becoming the dominant identification and tracking technology for objects and within facilities.

4.4.1 Radio Frequency Identification

The *automated authentication procedure* (*Auto-ID*) mainly used for the identification and tracking of objects is RFID. It offers a solution to the two mentioned main problems with bar-codes. RFID tags only need to get in the *proximity* of a scanner/reader in order to be read and can be made *tamper-proof*, if needed.

An RFID system is always made up of two main components:

Table 4.1 Road map of EPC transponder class types

EPC class type	Features	Transponder type
Class 0	Read-only	Passive
Class 1	Write once, Read many (WORM)	Passive
Class 2 (Gen. 2)	Read/Write	Passive
Class 3	Read/Write	Semi-active
Class 4	Read/Write	Active

1. the *transponder* or *tag*—placed on object or person to be identified; can be active or passive (i.e. can possess a source of energy or not)
2. the *reader* or *interrogator*—placed anywhere in the range; depending on the design and the technology used, they can act as read or write/read devices

RFID transponders can be classified in different ways. The first classification is characterized by their dependence on the energy from the reader. *Passive transponders* do not posses their own power supply. They receive all their energy from the reader through their antenna. They are typically used for object identification and tracking. *Active transponders* posses their own power supply in the form of a battery or solar cell. They act as beacons and their signals can reach readers even a few hundreds of meters away. They are typically used for persons or asset tracking. *Semi-active transponders* also posses their own power supply, however their behavior is different. They await signals from readers, before they are activated. Once interrogated they can provide instant access at distances of up to a 100 m. This prolongs their battery life, which may last up to 5 Years. They are typically used for asset, vehicle or yard management.

Further classifications of RFID transponders by frequency band, range and functionality is also possible. RFID systems operate at a wide range of frequencies, ranging from 135 kHz, up to 5.8 GHz. Operational ranges achievable by RFID tags vary from a few millimeters to a few hundreds of meters. Functionally, RFID tags are classified into 5 EPC groups (see Table 4.1).

RFID readers are set-ups of reader's antennas and back-end servers, where all identification data are collected. We can classify them by their realization into two groups: stationary and hand-held. Stationary can have more antennas and a wider range, whereas handhelds can be taken anywhere within the range of a corporate wireless communication network, interconnecting them with the back-end server.

4.4.2 RFID Security Mechanisms

RFID tags often contain sensitive information that can only be read by dedicated readers, paired with the tags for the sake of security (e.g. remotely unlocking the car and disabling the car ignition blockade). Hence, RFID security in its basic form isn't considered a novelty any more. Since, according to [18], in the near future we are going to pair ubiquitous sensing with RFID, weak RFID encryption will not suffice. The need for trusted readers shall arise, especially for ones that shall not only be solely authorized for reading certain groups of tags, but more importantly, ones that shall be solely allowed to add information to them.

For the sake of introducing security in RFID, several lightweight cryptographic protocols were developed. They can provide for basic security, but are easily breakable. Some of the most widely used of these protocols are *XOR*, *Subset*, *Squaring* and *Knapsack* [3]. In addition, one may combine different lightweight protocols to provide for additional security, use non-linearity as a set of pre-images, or use compression where authentication doesn't need to be based on inverse transformation. However, none of the aforementioned is suitable for broader use, because they are proprietary and don't provide for strong security.

The biggest problem in RFID security is the lack of computing power inside the transponder chip. The second biggest problem concerns passive transponders of class 0 and 1, which are read only (or can be written just once) and are not capable of any complex computing operations. Finally, even though one could overcome the lack of computing power, with RFID we are still dealing with wireless communication with a large variety of possible attacks: *eavesdropping*, *man-in-the-middle*, *Denial of Service (DoS)* and *replay attack*.

As protection against the latter, a variety of cryptographic solutions can be used, but none of them is "bullet-proof". Let us first consider *symmetric cryptography*. Since the same key is used for both—encrypting and decrypting datagrams—both sides need to exchange the secret key at some point in time, which is when they are vulnerable. *Triple DES* (*3DES*) encryption algorithm is a symmetric-key block cypher, which applies the *Data Encryption Standard* (*DES*) algorythm three times to each data block, using three different keys. It is often used to secure credit card payments, but more widely used to secure money transfers among banks in general. It was considered a strong and secure algorithm, but only among certified users. Since it is a symmetric cypher, in case a key is stolen, the communication is not secure any more [3, 19]. Due to its susceptibility to the man-in-the-middle attack, it is no longer a part of the Open SSL standard since version 1.1.0 (2016) and has been deprecated for general use by NIST in 2017. Another cryptographic option is to use *asymmetric cryptography*, but since all its known applications require quite a lot of computing power, strong security protocols to provide for e.g. *Pretty Good Privacy* (*PGP* [20–22]) in general cannot be applied to RFID.

To prevent man-in-the-middle and eavesdropping by design a new RFID technology has been developed, namely, *Near-Field Communication* (*NFC*). It is often built into RFID class 5 devices like smartphones to provide for secure bi-directional identification and communication at close distances. By its operating distance of up to 10 cm it rules out the two most stringent attacks mentioned, rendering also the third one hard to perform. Hence, nowadays it us commonly used with online payment systems applications such as Google and Samsung Pay. In combination with a smartphone's secure storage and biometric authentication, it may also be used enable the owner to store the intrinsic properties of his/her smartcards, like rail, credit and debet cards. Hereby, the users of such smartphones are enabled to identify themselves, and/or perform micro-payments, depending on the foreseen application of the scanned-in smartcards.

4.5 Synopsis

Computing devices of the future will interact with us in a more humanoid fashion [23]. A key element of that interaction will be their ability to recognize our faces and even understand our gestures and expressions. Being unobtrusive, face recognition has a natural place in the next generation smart environments. It takes place at a distance without a pause-and-declare interaction, in contrast to some other forms of biometric identification like fingerprint or iris scan. It does not restrict user movement, and it is now both—low power and inexpensive.

Face recognition systems shall no longer be limited to identity verification and surveillance tasks. Their applications will be able to interpret human actions, intentions, and behavior as central part of the next generation smart environments. This will have a great impact on the whole community regarding safety and security. A good example thereof is the early detection of potential terrorist attacks by behavioral monitoring in crowded public places. Detecting behavior in combination with face recognition could be the answer to prevent various criminal actions. In such applications, identity information is crucial to providing machines with the background knowledge needed to interpret biometric measurements and observations of human actions.

A single biometric sensor may sometimes fail to be extract enough distinctive features for a person's identification (for example identical twins—their faces alone may not distinguish them) [24]. So smart cities shall need the infrastructure and applications for double-checking identities of individuals entering or accessing secure places. By combining face, iris, voice, fingerprint, lip movement recognition and other security measures one would expect such systems would to become impenetrable. All modalities combined should result in a safer community for all residents of a smart city.

According to [25] the ability to uniquely identify individual objects is essential to many aspects of modern life such as manufacturing, distribution logistics, access control, and fighting terrorism. Radio frequency identification (RFID) is a wireless communication technology that has proven useful for precisely identifying objects. RFID uses radio-frequency waves to transfer identifying information between tagged objects and readers without the need for a line of sight (LOS), providing a means of automatic identification.

With RFID [25] it is possible to read encoded information even if the tag is concealed for either aesthetic or security reasons—for example, embedded in product's casing, sewn into clothing, or sandwiched between banknote's layered papers. Such enhanced RFID tags shall have a big influence on logistics in smart green cities in terms of more efficient supply-chain management, where RFID would help close information gaps by enabling real-time supply chain visibility. Also, it shall bring great progress to healthcare and pharmaceuticals industry where RFID technology shall reduce operating costs and improve patient safety in hospitals by providing for traceability of patients, doctors, medical assets, and medicines.

With all of this in mind the mechanisms to provide for strong RFID and biometric security have been investigated targeting sensor networks applications in logistics (cp. Chap. 14) and healthcare (cp. Chap. 5 on e-health and Chap. 15 on telediagnostics).

References

1. Geer D (2006) Security of critical control systems sparks concern. Computer 39(1):20–23. https://doi.org/10.1109/MC.2006.32
2. Hole KJ, Dyrnes E, Thorsheim P (2005) Securing wi-fi networks. Computer 38(7):28–34. https://doi.org/10.1109/MC.2005.241
3. Kaps JP, Gaubatz G, Sunar B (2007) Cryptography on a speck of dust. Computer 40(2):38–44. https://doi.org/10.1109/MC.2007.52
4. Pankanti S, Bolle RM, Jain A (2000) Biometrics: the future of identification [guest editors' introduction]. Computer 33(2):46–49. https://doi.org/10.1109/2.820038
5. Phillips PJ, Martin A, Wilson CL, Przybocki M (2000) An introduction evaluating biometric systems. Computer 33(2):56–63. https://doi.org/10.1109/2.820040
6. Phillips P, Martin A, Wilson C, Przybocki M (2000) An introduction to evaluating biometric systems. Computer 33:46–50
7. Wang K, Ma H, Popoola O, Li J (2011) Biometrics: finger vein recognition. In Tech, Croatia
8. Bazen A, Verwaaijen G, Gerez S, Veelenturf L, Van der Zwaag B (2000) A correlation-based fingerprint verification system. In: Workshop circuits systems and signal processing
9. Feng J, Ouyang Z, Cai A (2006) Fingerprint matching using ridges. Pattern Recognit 39:2131–2140
10. Ratha N, Bolle R, Pandit V, Vaish V (2000) Robust fingerprint authentication using local structural similarity. In: Fifth IEEE workshop applications of computer vision
11. Bazen A, Gerez S (2003) Fingerprint matching by thin-plate spline modelling of elastic deformations. Pattern Recognit 36:1859–1867
12. Feng J, Jain AK (2011) Fingerprint reconstruction: from minutiae to phase. IEEE Trans Pattern Anal Mach Intell
13. Agopov M (2011) Biometrics: retinal identification. In Tech
14. Richard A, Larsson S (2009) Genetics of human iris colour and patterns pigment cell melanoma 22:544–562
15. ISO: Iso/iec 16022:2006 information technology. https://www.iso.org/standard/44230.html
16. Finkenzeller K, Müller D (2010) RFID handbook: fundamentals and applications in contactless smart cards, radio frequency identification and near-field communication, 3th edn. Wiley
17. Xiaolin J, Quanyuan F, Taihua F, Quanshui L (2012) RFID technology and its applications in internet of things (IoT). In: Jia X, Feng Q, Fan T, Lei Q, pp 1282–1285
18. Want R (2004) Enabling ubiquitous sensing with RFID. Computer 37(4):84–86. https://doi.org/10.1109/MC.2004.1297315
19. McLoone M, Robshaw M (2006) Public key cryptography and RFID tags. In: Abe M (ed) Topics in cryptology—CT-RSA 2007, vol 4377. Lecture notes in computer science. Springer, Berlin Heidelberg, pp 372–384

20. Batina L, Guajardo J, Kerins T, Mentens N, Tuyls P, Verbauwhede I (2007) Public-key cryptography for RFID-tags. In: IEEE international conference on pervasive computing and communications workshops, pp 217–222
21. Peris-Lopez P, Hernandez-Castro JC, Tapiador JM, Ribagorda A (2009) Advances in ultra-lightweight cryptography for low-cost RFID tags: gossamer protocol. In: Chung KI, Sohn K, Yung M (eds) Information security applications, vol 5379. Lecture notes in computer science. Springer, Berlin Heidelberg, pp 56–68
22. Bilal Z, Masood A, Kausar F (2009) Security analysis of ultra-lightweight cryptographic protocol for low-cost rfid tags: gossamer protocol. In: International conference on network-based information systems. NBIS '09, pp 260 –267
23. Pentland A, Choudhury T (2000) Face recognition for smart environments. Computer 33(2):50–55. https://doi.org/10.1109/2.820039
24. Frischholz RW, Dieckmann U (2000) Biold: a multimodal biometric identification system. Computer 33(2):64–68. https://doi.org/10.1109/2.820041
25. Li X, Sheng QZ, Zeadally S (2008) Enabling next-generation RFID applications: solutions and challenges. Computer **41**:21–28. https://doi.ieeecomputersociety.org/, https://doi.org/10.1109/MC.2008.386

Chapter 5
E-Health

By the aging of the world population, increase of chronic diseases, increasing people's demands for new, more complex diagnostic and therapeutic methods, as well as current distribution and capacities of health providers, the way towards the introduction of new *health services* is being paved. They shall be based on new process models and advanced information and telecommunications solutions. *Telediagnostics* and *telemedicine* are promising approaches to better health services of the future, being more effective than existing, established models of healthcare.

5.1 Introduction

By applying the *telehealth* (*e-health*) approach, a more efficient way of equally personalized but more timely multidisciplinary specialist treatment can be introduced. However, by its introduction also new data security issues in healthcare systems shall arise. While on one hand, data in electronic forms are easy to access, track and archive, and they may also travel very fast, on the other, these properties are also facilitating their easier abuse.

While ensuring *patient safety* is primarily an organizational concern, in telehealth one should also consider *patient information security* and make use of the mechanisms provided by the enabling (information) technology. In telehealth proper information security management should provide for *confidentiality*, *integrity*, and *availability* (CIA) of patient's *Electronic Medical Records* (*EMR*). Patient information security should be an integral part of *telehealth services*, since the healthcare professionals involved are morally, ethically, and legally responsible for their patients' medical records [1].

R. Gumzej, *Intelligent Logistics Systems for Smart Cities and Communities*,
Lecture Notes in Intelligent Transportation and Infrastructure,
https://doi.org/10.1007/978-3-030-81203-4_5

While the terms telehealth, telediagnostics, and telemedicine are often used interchangeably it is important to note that *telehealth* is the all-encompassing term, while *telediagnostics* and *telemedicine* pertain to provision of diagnostic and therapeutic treatments on a distance by the use of information and communication technology.

5.2 Telehealth Applications

The constraints of the healthcare systems in developing countries, including high population growth, high burden of disease prevalence, lacking healthcare personnel, large population of rural inhabitants, and limited financial resources to support healthcare infrastructure, have motivated the development of *telehealth* (*e-health*) and *mobile health* (*m-health*) applications. M-health is known as the practice of medical and public health support to delivering telemedicine services through mobile devices such as smartphones and tablets. Thus, the increasing popularity of m-health can be associated with the rapid development of wearable (medical) devices and wireless communication technologies.

To fully utilize smart mobile, wearable and wireless technologies, the concept of a *Body Sensor Network* (*BSN*), being a kind of a wearable *Wireless Sensor Network*, was proposed in 2002 [2]. The concept of the wireless juxta-corporal sensor-net on a human body, may represent a basis of a typical setup for e-health and m-health (Fig. 5.1) applications. The network consists of diagnostic *sensor nodes* that gather data on the physical state of a patient, being transferred to his/her *Electronic Medical Record* at a hospital's *Medical Server*, and *Terminals/control nodes* accessing this data, representing terminals owned by medical personnel, hospital's administrative personnel, insurance company clerks, etc.

The BSN concept has great potential as a prototype front-end platform for telehealth applications. Consisting of sensors attached to the human body for collecting and transmitting patient's blood pressure, saturation, ECG, EEG, and similar sensor data, the BSN is able to facilitate the joint processing of spatially and temporally organized medical data, resource optimization and systematic health monitoring.

Figure 5.1 presents a simplified example of a BSN application scenario in an m-health system. *Sensor nodes* are joined with a *Master Node* (*MN*) to form a BSN, representing a patient entity. Medical information collected from different sensors on a patient's body are being sent to the MN for data fusion and transmission to a personal (mobile) terminal for prepossessing before being forwarded to a medical server and/or a physician's terminal for diagnostic purposes and/or telemedical treatment. In response to the provided diagnostic data, the physician would provide appropriate guidance to the patient or his/her medical assistant through the patient's (mobile) terminal for proper treatment or medication. In case of an urgent medical state, the physician would provide the patient with appropriate instructions to stabilize his/her condition and organize an emergency response medical team to assist the patient on site. Besides teleconsultation this is a typical e-health/m-health scenario in which the BSN would play a crucial role for providing health services to patients in environments with limited access to medical facilities.

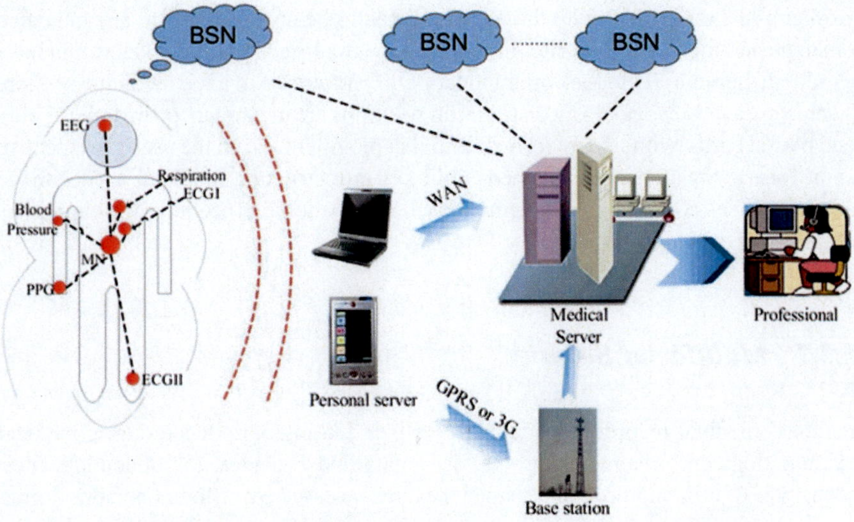

Fig. 5.1 M-health application scenario [2]

5.3 Telehealth Safety and Security

Wireless networks for medical applications have to fulfill a high level of reliability to guarantee the safety and security of telehealth applications as well as the privacy of patients' healthcare records. To ensure the security of the overall e-health system, BSNs as its front end, should be protected from different possible attacks such as eavesdropping, injection and modification. However, this is a non-trivial task due to stringently limited processing, memory, and energy capacity as well as the lacking appropriate user interfaces for non-expert users, longevity of devices, and global roaming capabilities for most sensor nodes [2].

Symmetric cryptography, in which all communicating parties must share a secret key via an invulnerable key distribution protocol prior to any information interchange involving the encryption process, is a promising approach to resolve the stringent resource constraints of BSN devices. Existing key distribution techniques for large-scale sensor networks, such as random-key pre-distribution protocols and polynomial pool-based key distribution [3], all require some form of pre-deployment. However, presuming an increasing number of BSNs being deployed, this approach may involve considerable latencies during BSN networks initialization and makes any subsequent maintenance works harder. In addition, it doesn't appropriately support two most obvious working scenarios. It discourages family members, to share the BSN because each one of them would need initialize the network with its own private ID. Whenever there should be a need to add or change a body sensor, the user would have to reconfigure the BSN to ensure that the new sensor can securely communicate with the existing ones. A common working scenario in telemedicine is also to have a health worker assist the patient with the use of a BSN, to be able to perform any indicated or suggested medication on site. In this case the qualified health worker would have to add his/her own ID to the communication to participate in a telemedicine session. Therefore, a new series of key distribution solutions are aspired to, providing for plug and play security without any form of initial deployment. From the user's perspective some form of the already established public key infrastructure should be used with the addition of session keys to supplement each telemedicine procedure, which implies *asymmetric cryptography*.

5.3.1 Multifactor Security

Patients' medical records, especially diagnostic findings, produced during medical examinations and treatments, are sensitive data and considered confidential. Their security is of critical importance, since they are accessed by different persons during medical treatment and, in part, managed by persons not involved in medical therapy.

On one hand, the introduction of electronic medical records has obvious benefits for the patients, their doctors as well as medical insurance companies due to the ease of access and handling. On the other hand, unauthorized accesses to diagnostic reports, correspondence, prescriptions, medical history, and insurance data is possible while accessing them, due to poor protection of the terminals through which they are accessed, as well as during transfer, since these data are often communicated between patients, doctors, hospital administration and insurance companies without using proper encryption.

To ensure secure access and confidentiality of medical data, especially the following precautions are considered crucial:

- Medical data should always be transferred through secure channels.
- The access to medical data should be protected by appropriate authentication mechanisms (passwords, Biometrics etc.).
- Patient's medical records should be verifiable by the patient in order to detect and correct possible inconsistent data on diagnoses, treatments or prescribed medications, to prevent negative consequences (e.g. ill-treatment and/or ill-medication, changes in medical insurance rate, professional disability etc.).
- Patients and medical personnel data should be appropriately protected from being misused.

In telehealth applications the *biometric approach* to authentication is appealing because of its convenience and possibility to provide for strong security with non-repudiation. Often the main obstacle to its broader use is the fact that specific hardware such as *biometric scanners* and complex software for *features extraction* and *biometric template matching* are required [4, 5]. However, since in medical environments the required equipment often is already present and partly being used for diagnostic purposes, the approach can be employed for securing *hospital information systems*, *patient's* EMR as well as access to sensitive *electronic diagnostic devices*.

Table 5.1 Authentication methods in telehealth

Authentication category	Methods
Non-biometric (memorized or possessed)	password, personal identification number, pass phrase, mobile device identification number, tokens (key fob), dongle, smart card, radio-frequency identification
Biometric (scanning patterns of the…)	face, fingerprint, hand, iris scan, retina scan, voice print, palm

For patients, physicians, medical staff and administrative personnel its use is both secure and comfortable, because they always carry their biometric characteristics with them—they cannot lose them, and abuse is almost impossible to perform.

A *data security solution* must include standardized policies, technologies, and administrative practices to assure data security. According to [6], a successful move towards a standardized data security methodology requires partnerships among consumers, private industry, advocacy groups and governments. The American Telemedicine Association has provided high-level guidance to providing for *m-health security* [7, 8] (cp. Table 5.1). Often two or more of them are used in conjunction.

Multi-factor authentication (*MFA*) is an electronic authentication method in which a computer user is granted access to a website or application only after successfully presenting two or more pieces of evidence (or factors) to an authentication mechanism: *knowledge* (something the user knows), *possession* (something the user has), and *inherence* (something the user is). Being a sound security solution, multi-factor authentication has also found its way to a broader area of user-oriented security applications, such as *e-governance* and *e-banking*. In summary, MFA may effectively protect the users in telehealth applications from unauthorized access to sensitive medical data and equipment.

5.3.2 Security of Body Sensor Networks

As BSNs already consist of nodes containing biosensors for collecting and processing patient's biometric data, it is also possible to use them for securing BSNs in e-health and m-health applications. It is well known that the human body physiologically and biologically consists of its own transmission systems such as the blood circulation, neural, digestion systems. Thus, making use of these unique physiological characteristics to provide for patient authentication is deemed quite practical [9]. For obvious reasons it also necessary to ensure that the sensors of the applied BSN are used on a single individual. Hence, two-stage authentication is necessary in order to provide for safety and security of a BSN application.

The use of physiological signals to identify individuals, representing the second phase of authentication, and transmitting their own encryption key securely within

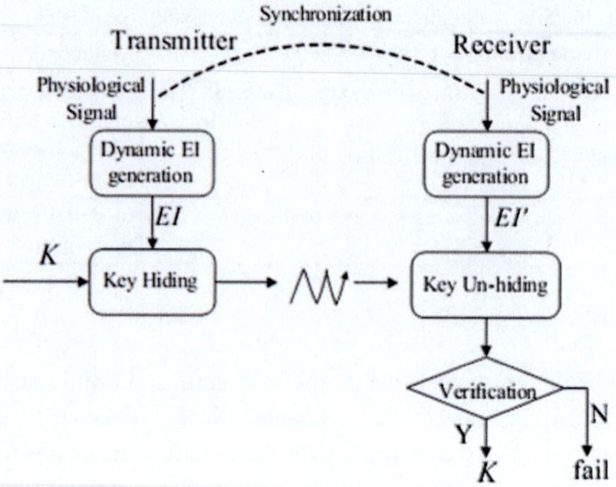

Fig. 5.2 Biometrics-based key generation and propagation within a BSN [2]

a BSN, representing the first phase of authentication, is considered feasible and can save resources. Building upon this initial assumption, a family of lightweight and resource-efficient biometric security solutions, being based on time-variant physiological signals, has been proposed. In an m-health application (cp. Fig. 5.1), the physiological signals of a human body, such as *electrocardiograph* (*ECG*) and *photoplethysmograph* (*PPG*), can be used to generate an *entity identifier* (*EI*) of each BSN (cp. Fig. 5.2) for patient's identification as well as key sharing within the BSN by a key hiding/unhiding processes. This authentication procedure is based on the fact that EIs generated simultaneously from the same subject share a great similarity, while those generated non-simultaneously or from different subjects exhibit significant differentiation. Hereby, both goals of the two-stage authentication process are achieved – secure key transmission within a BSN and patient's identification and authentication with an m-health application. As opposed to traditional Biometrics, where the physiological or behavioral characteristics are static and merely used to automatically identify or authenticate an individual, the physiological characteristics in a BSN are dynamic and as such can assure a higher degree of security [2].

5.4 Synopsis

Biometrics can establish personal identities from the moment patients enter a medical facility for the first time, by scanning in their *biometric signatures*. This identity can subsequently be used as part of his/her *digital certificate* to safely and securely transmit patient's data throughout the *healthcare information system*.

Biometrics can be used to ensure that only authorized medical personnel can access sensitive patient's diagnostic data as well as hospital facilities, such as pharmacy, nursery and operating rooms, to see to it that prescribed medications are properly delivered to the patients, and to safeguard the privacy of patients' medical records by assuring that only authorized personnel can access them. Since in medical facilities devices to perform biometric scanning are already present for diagnostic purposes, Biometrics may be considered the future of positive healthcare identification and will enhance the secure use, storage, and exchange of patient's medical records. Hence, in telehealth Biometrics may be considered a part of the critical infrastructure.

Multifactor authentication and authorization is considered a contemporary method of protecting medical data stored in healthcare information systems. Since biometric characteristics are unchangeable and non-detachable, it is impossible to lose them and misuse them as such. Hence, biometric user authentication in public services, especially in health services, is considered both secure and effective. In combination with the patient/personnel digital certificates, electronic diagnostic devices shall enhance the *confidentiality*, *integrity*, and *availability* of *medical data*. By safe and secure telehealth applications, the patients would gain trust in personalized telediagnostics and telemedicine applications (cp. Chap. 15) and the notion of a better healthcare service due to shorter waiting times and reduced need to travel to obtain specialist treatment.

References

1. Harrington L, Kennerly D, Johnson C (2011) Safety issues related to the electronic medical record (EMR): synthesis of the literature from the last decade, 2000-2009 **56**(1):31–44. https://doi.org/10.1097/00115514-201101000-00006
2. Miao F, Bao S, Li Y (2012) New trends and developments in biometrics: physiological signal based biometrics for securing body sensor network. In Tech
3. Liu D, Ning P (2003) Establishing pairwise keys in distributed sensor networks. In: the 10th ACM conference on computer and communication
4. Le C A survey of biometrics security systems. https://www.cse.wustl.edu/~jain/cse571-11/ftp/biomet/
5. Yang J, Xie S (2012) New trends and developments in biometrics. IntechOpen. https://books.google.si/books?id=DG7VyAEACAAJ
6. Luxton D, Kayl RA, Mishkind MC (2012) mhealth data security: the need for hipaa-compliant standardization. Telemed J E-Health: Off J Am Telemed Assoc 18(4):284–8
7. Health & Human Services UD Health information privacy. https://www.hhs.gov/hipaa/for-professionals/index.html
8. Yellowlees P, Shore J, Roberts L (2010) Practice guidelines for videoconferencing-based telemental health—October 2009. Telemed J E-Health: Off J Am Telemed Assoc 16:1074–89. https://doi.org/10.1089/tmj.2010.0148
9. Poon CC, Zhang Y, Bao SD (2006) A novel biometrics method to secure wireless body area sensor networks for telemedicine and m-health. IEEE Commun Mag 73–81

Chapter 6
E-Commerce

According to [1], a *supply chain* (*SC*) is the network of organizations that are involved, through upstream and downstream linkages, in the different processes and activities that produce value in the form of products and services in the hands of the ultimate consumer. For example a clothing manufacturer is a part of a supply chain that extends upstream through the weavers of fabrics to the manufacturers of fibers and downstream through the distributors and retailers to the final consumers. Each of these organizations in the chain are dependent on each other by definition and yet, paradoxically, by tradition do not closely co-operate with each other.

Traditionally, most organizations have considered themselves as entities that exist independently from others and need to compete with them in order to prosper. However, this attitude can be self-defeating, if it leads to unwillingness to co-operate in order to compete. The background of this philosophy is the basic idea of *supply chain integration*.

6.1 Introduction

According to the http://www.apics.org/dictionary/dictionary-information?ID=3984 APICS Dictionary, *Supply Chain Management* (*SCM*) comprises the design, planning, execution, control, and monitoring of supply chain activities with the objective of creating net value, building a competitive infrastructure, leveraging worldwide logistics, synchronizing supply with demand, and measuring performance globally.

Managing a supply chain involves the flow of both tangible and intangible resources including materials, information, and capital across the entire supply chain. The flow of these resources can be unidirectional (e.g., services) as well as bidirectional (e.g., materials, products, capital, and information). A supply chain can contain from two to several stages (e.g., suppliers, manufacturers, distributors, retailers) and

R. Gumzej, *Intelligent Logistics Systems for Smart Cities and Communities*,
Lecture Notes in Intelligent Transportation and Infrastructure,
https://doi.org/10.1007/978-3-030-81203-4_6

each of these stages can possibly involve several entities (e.g., production sites, storage facilities, etc.) that either compete or cooperate to provide for the service, product, or materials requested from a further stage in the chain.

Supply chain management has gained a huge interest among researchers in the past few years. This was mainly to ensure timely information flow across different stages of the supply chain and also to effectively utilize these information for improved performance. Although information in supply chains plays a significant role, the dynamics of supply chains and the ways the data in these systems are collected are gaining on importance.

Intelligent Logistics Systems and *Augmented Logistics* (cp. Chap. 8) represent a possibility of *supply chain automation* to cope with the increasing dynamics in the global *electronic marketplaces* (*e-marketplace*s).

6.2 E-Marketplaces

A common classification of *business-to-business* (*B2B*) electronic marketplaces distinguishes between two basic types of e-marketplaces:

1. *horizontal e-marketplaces* support cross-industry functions and processes, for example by offering different value added goods and services indirectly, not specializing on any specific line of industry, based on a one-stop-shop principle; e.g.: amazon.com, walmart.com.
2. *vertical e-marketplaces* support businesses along the value chain of a certain line of industry and are thus targeted at specific customers or service areas (e.g. food, textile, chemical industry). Since the traded products and services address the core businesses of companies in the chain, vertical e-marketplaces are more important for *inter-organizational cooperation* than on horizontal e-marketplaces; e.g.: covisint.com, SupplyOn.com, Elemica.com.

E-marketplace systems can be further categorized as open and closed. While *open e-marketplaces* are available to all users, *closed e-marketplaces* are exclusively available only for selected participants, e.g. a certain industry, product line, certification class, etc.

The services, offered in e-marketplaces, which are often used for further classification, include: bulletin-boards, catalog-based services, exchanges and auctions. A common service based classification of B2B e-marketplaces concerns the fact whether they support *Order-to-Invoice* (*O2I*) or *Vendor Managed Inventory* (*VMI*) processes. The key difference between these two approaches lies in the distinctive property of the latter. Namely, by VMI a supplier is given insight into the customer's business operations and their production and demand plans, whereas following the O2I policy, the supplier has no insight into what is going to come next. On one side, VMI represents a great responsibility for a supplier, whereas on the other it is also a great source of market intelligence. It allows the supplier to plan ahead and avoid last-minute rush orders to meet unexpected excess orders from its customers. Furthermore, suppliers can draw down their inventories, since there is no need to keep "just-in-case" inventories on hand in case their customers would unexpectedly order a large volume of a certain material.

So, how to take B2B Integration to the next level? By managing complexity with B2B integration processes. One cannot do business or run a supply chain without some form of integration. Since integration means different things to different people, there are many different approaches to B2B integration. As a result, most companies are dealing with tremendous complexity in managing their B2B processes. In process industries (e.g., chemical, pharmaceutical, and food processing) it is clear that there is a significant difference in capabilities when companies use B2B Networks versus EDI, fax, email, or spreadsheets. The latter introduce too much latency, and are not equal to the challenge.

For example, SAP Commerce is an open cloud based solution for horizontal, vertical, and personal e-marketplaces. The platform fulfills the function of a router of business documents between marketplace-external and -internal systems. The SAP Commerce Marketplace connectors, called accelerators, exceed the functionality of the SAP BC (Business Connector), by which EDI-based seamless integration of enterprise IT solutions with the mySAP Ecosystem was enabled in the past. These components include the SAP application areas Connectivity, Security and Management. Similar to the SAP Commerce solution, the ERP systems of contributing companies may be interconnected with other companies through the Elemica network to create B2B supply chain connections. Hereby, Elemica creates an abstraction layer for interconnections between the various IT solutions used by various companies. Thus, e-marketplace automation can be implemented without the need to change the business processes of any participating company.

6.3 Supply Chain Operating Networks

While B2B integration has changed substantially during the last decade, a couple of clear trends have been identified—e-marketplaces are becoming (1) more accessible and (2) easier to do business with. Both can be achieved with what is called the next level in B2B integration, such as *Supply Chain Operating Networks* (*SCON*), addressing agility with market-driven supply chains. Supply Chain Operating Networks bring together trading partner connectivity with *Software-as-a-Service* (*SaaS*) middleware applications. Instead of companies creating hundreds or thousands of diverse one-to-one connections with their trading partners, they make a single connection to the business network, where their trading partners and thousands of other companies are also connected, and they use the SaaS applications that reside on the network to communicate, collaborate, and execute business processes in more efficient, scalable, and innovative ways.

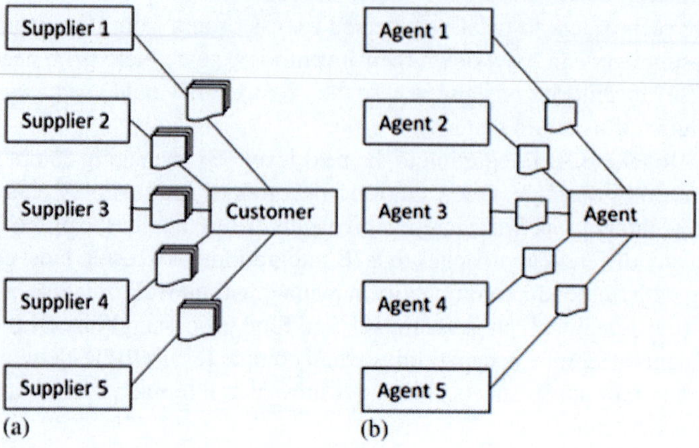

Fig. 6.1 A supply chain before **a** and after **b** SCON integration

Figure 6.1a presents a traditional e-marketplace in which B2B transactions are being performed in various forms and protocols (EDI, e-mail, Fax,...). In Fig. 6.1b B2B transactions are being performed within a SCON in the form and protocols, determined by the e-marketplace.

Generally, two approaches to service transformation in the digital era can be observed for efficient, service-quality-oriented *Supply Chain Management*:

1. *Centralized e-marketplaces*, combining supply and demand in temporally and spatially shared repositories of SCON operators (e.g., SAP Business Network, Elemica.com) and
2. *Distributed e-marketplaces*, based on multi-agent systems with a yellow-page dictionary service and a network of temporally and spatially distributed agents of supply chain nodes (e.g., Chap. 17).

While centralized supply chain operating networks are well suited for vertical e-marketplaces, horizontal e-marketplaces are harder to tackle. By the introduction of decentralized P2P e-marketplaces, realizing the Supply Chain Operating Networks with *autonomous agents*, an unlimited universal e-marketplace framework with all the listed attributes can be realized.

6.4 Real-Time Ability

Real-time describes various operations in computing or other processes that must guarantee response times within a specified time (deadline), usually a relatively short time. A *real-time process* is generally one that happens in defined time intervals of maximum duration and fast enough to predictably affect the environment in which it occurs, such as a response to input signals from a computing system. *Real-time ability* is also an important property of business transactions. Most of them have some temporal restrictions, which determine their validity. Eventually, this influences their *timeliness*, which is a *service quality criterion*.

Intra-organizational real-time ability is defined by the zero-latency enterprise concept. *Zero-latency enterprise* means that any change in an enterprise information system is evident in any part of the information system without noticeable delay. Within e-marketplaces and supply chains this concept is translated into the concept of *available-to-promise* (*ATP*) time. Available-to-promise is a business function that provides a response to customer order inquiries, based on resource availability. It relates to timeliness in the sense of predicting the ultimate point in time (*deadline*), when a resource or material would be available. It has been derived from the push and pull ordering schemes in supply chain management dealing with forward or backward scheduling to determine that point in time. While these two concepts are crucial for achieving the synchronicity of physical and data flows, there is a third notion, which is specific to e-marketplaces.

Timeliness in e-marketplaces should be defined as the concept assuring *consistency* and *correctness* of transactions performed on-line, regardless of the type of industry and realization of the e-marketplace system. Since supply chain management operations depend on the validity of data within the e-marketplace system, their consistency and correctness in space-time is the source of trust in supply chains they manage. Hence, they are also considered their *service quality criterions*.

6.5 Synopsis

In this chapter the most important aspects of e-commerce digitization have been dealt with. While the digitization of intra-organizational processes can be realized by ERP systems, the digitization of inter-organizational cooperation is best realized by e-marketplaces. Due to the otherwise necessary adapter components for the various B2B transactions among business partners and still lacking service quality in terms of consistency, correctness and timeliness, the introduction of e-marketplaces is seen as the ultimate contemporary supply chain management solution.

Depending on the nature of businesses, companies can decide between horizontal and vertical supply chain connectivity and do so in a centralized or decentralized manner. Their decision impacts the visibility of their business in the e-marketplaces of choice, but nevertheless adds to their global recognition. The stated differences in the realization of e-marketplaces can be followed through towards their maintainability and flexibility—i.e. sustainability. In the long run decentralized e-marketplaces are more sustainable, since they naturally correlate to the global economy. They tend to be harder to realize, but are easier to maintain, since they initially overcome boundaries which in time tend to become obstacles of centralized e-marketplaces, like geographical boundaries and restrictions to certain lines of industry.

Finally, a word on the timeliness topic, last addressed in this chapter. It is as important for the synchronicity of the physical, monetary and data flows as it is vital for the correct and consistent execution of any supply chain management transaction. Hence, it should be taken very seriously. An example of a time-critical application is *High Frequency Trading* [2]. Without the real-time ability, high-frequency trading, being one of the most important drivers in today's stock and monetary markets, would not be possible.

Good examples of a both centralized and decentralized e-marketplaces are described in Chaps. 16 and 17. Here, the above mentioned service quality criteria are already incorporated into the business logic of e-marketplace automation. Through the following chapters the necessity of this design decision will become even more evident.

References

1. Christopher M (1992) Logistics and supply chain management strategies for reducing costs and improving service. Pitman Publishing, London
2. Bundesbank D (2016) Bedeutung und wirkung des hochfrequenzhandels am deutschen kapitalmarkt. https://www.bundesbank.de/resource/blob/665078/544876d8a09dd548ed15bd74ce14281f/mL/2016-10-hochfrequenzhandel-data.pdf

Chapter 7
Industry 4.0

The initiatives "Industry 4.0" (Germany) and "Made in China 2025" (China) stand for smart production of smart products. Enabling the realizations of these concepts, the Intelligent Web (Web 2.0) in general and the Internet of Things (IoT) in particular are needed as infrastructural basis.

7.1 Introduction

Intelligent production and transportation systems are utilized in production, services, storage facilities and transport. Characteristic to them is the use of information technology, steering their automated activities or the actions of their users to enable their autonomous operation. Since these systems and their subsystems and components are accessible and/or obtain their knowledge from the Web, they are often termed *Cyber Physical Systems* (*CPS*).

In a broad sense, anything "smart" needs to be implemented as a *Cyber Physical System* (CPS). As cooperative information systems they are the enablers of the future Society 5.0. Their basic properties correlate with the properties of *autonomous systems*: *distributedness*, *autonomous operation*, *self-configuration* and *self-learning*. In order to function as such, they need to provide for diverse and "open" interfaces, safe execution environments and the ability to participate in various kinds of communication networks. Depending on their environment they may have more or less strict temporal restrictions to their operation, but in any case they need to operate in *real-time*.

R. Gumzej, *Intelligent Logistics Systems for Smart Cities and Communities*,
Lecture Notes in Intelligent Transportation and Infrastructure,
https://doi.org/10.1007/978-3-030-81203-4_7

7.2 Intelligent Production Systems

When using *Intelligent Production Systems* (*IPS*) one relies on their functional *safety* as well as *security* of their stored and transferred data, which need to be assured according to appropriate standards on all levels of their operation. Appropriate standards are under development and address the safety and security of CPS or *Industrial Control Systems* (*ICS*), as their most important subset.

The *dependability* of ICS is usually assured by redundancy and over-scaled components. This results in more complex designs and higher costs, but often without guaranteeing safety or security. To achieve their better overall quality, much effort was invested in search for standardized components, methods and tools apt to improve designed system's *predictability* and *dependability*.

The design and development methods of contemporary ICS are well established, relatively cheap and widely used. Hardware components come with specifications, which undoubtedly state their capabilities and performance indicators. Complexity increases, however, when there is a need for their integration into larger set-ups and system-level performance must be assured. Software makes things even more complicated, as the "*Write Once Run Everywhere*" (*WORE*) principle is still hard to achieve, and different software engineering techniques can lead to application programs with very different quality (of service), running on the same hardware platform. To achieve a managed level of quality (of service), systems engineering methods should enable *hardware-software co-design* (*HW/SW Co-design*) as well as efficient system's *prototyping, verification and validation* (*V&V*) before putting them to use.

In summary, *Intelligent Production Systems* need to be designed holistically utilizing the *systems approach* to their *engineering* with respect to appropriate *guidelines* and *standards*, as presented in [1]. In their design and development *model based techniques* should be used, combining the models of their *software*, *hardware*, and *netware* components.

7.3 ANSI/ISA-95

A number of formats have been developed for *interoperability* and information sharing among enterprise information systems. To be able to accumulate, manage and distribute information from and to various components, a true open source and comprehensive (stack of) standard(s) are required for knowledge management and utilization. They shall form the foundation of intelligent production systems.

The first step in this direction was the introduction of the ANSI/ISA-95 standard. It is an international standard of the International Society of Automation for developing automated interfaces between enterprise and control systems. This standard has been developed for global manufacturers to be applied in all industries, and in all sorts of processes, like batch processes, continuous and repetitive processes.

Its objectives are to provide a consistent terminology as a foundation for supplier and manufacturer communications, to provide consistent information models, and to provide consistent operations models, being a foundation for clarifying application functionality and how information is to be used. It contains terminology and models that can be used to analyze any production system: *functional model* (classification of business functions), *equipment model* (classification of business units and production devices), *information interchange model* (data flows in business processes).

ANSI/ISA-95 consists of five parts:

1. ANSI/ISA-95.00.01-2010 (IEC 62264-1 Mod) Enterprise-Control System Integration—Part 1: Models and Terminology
2. ANSI/ISA-95.00.02-2010 (IEC 62264-2 Mod) Enterprise-Control System Integration—Part 2: Object Model Attributes
3. ANSI/ISA-95.00.03-2013 Enterprise-Control System Integration—Part 3: Activity Models of Manufacturing Operations Management
4. ANSI/ISA-95.00.04-2012 Enterprise-Control System Integration—Part 4: Objects and attributes for manufacturing operations management integration
5. ANSI/ISA-95.00.05-2018 Enterprise-Control System Integration—Part 5: Business-to-Manufacturing Transactions

It has mainly been used in the realizations of *Manufacturing Execution Systems* (*MES*), whose main role is to close the gap between the managerial *Enterprise Resource Planning* (*ERP*) and *Process Control Systems* (*PCS*) in *automated production systems*. By their functionality MES systems mainly provide for transparency in resource management and optimization of production processes, and may improve production systems *flexibility* and *maintainability*. Due to a large diversity of different production processes, materials and tools, a unification standard like the ANSI/ISA-95 was necessary to be able to implement MES systems. By their implementation further goals could be achieved—improved *quality management system* due to increased overall production process transparency, as well as *licenseability* and *interoperability* of products and business processes due to the introduction of common standards. ANSI/ISA-95 effectively addresses the diversity of *enterprise control systems*.

7.4 OPC UA

As opposed to the diversity of enterprise control systems, a similar diversity of communication infrastructure in industrial environments can be observed. *Industrial Ethernet* has established itself as a industry standard for intra-organizational information interchange in production environments. Some major proprietary Industrial Ethernet protocols in use include: EtherCAT, EtherNet/IP, PROFINET, POWERLINK, SERCOS III, CC-Link IE, and Modbus TCP. On the application (Business Connector) level XML and JSON *electronic data interchange* (EDI) standards like the *ANSI ASC X12* have established themselves both for intra- and inter-organizational information

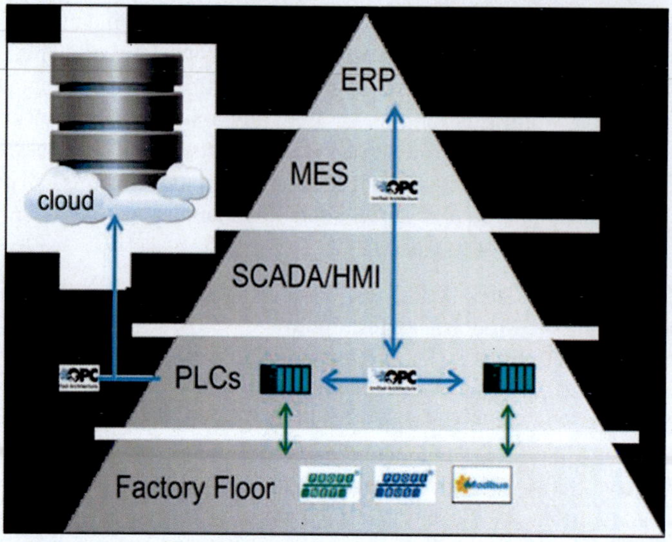

Fig. 7.1 Vertical and horizontal integration with OPC Unified Architecture in factory automation [2]

interchange. Due to proprietary nature of most of this standards, large scale interconnections as well as interoperability among businesses have been hard to achieve. To be able to control the devices through the Intra- as well as the Internet, the need for open source and comprehensive unification standards arose here as well.

Open Platform Communications (OPC) is a series of standards and specifications for industrial telecommunication. OPC specifies the communication of real-time plant data between control devices from different manufacturers. Being an open-source collaboration standard it is increasingly gaining importance. It facilitates interoperability of production systems and devices and provides for open and standardized information interchange between the managerial and process control levels (Fig. 7.1).

OPC Unified Architecture (OPC UA) is a vendor-independent communication protocol for industrial automation applications. OPC UA is a machine to machine communication protocol for industrial automation developed by the OPC Foundation. It is based on the client-server principle and allows for seamless integration and communication between the individual sensors and actuators on the shop-floor level up to the company's ERP systems and even cloud based enterprise information systems. Its protocols are platform-independent and feature built-in safety mechanisms. OPC UA introduces binary and Web interfaces in order to provide for transparency and accessibility of information and services. It represents the foundation of the foreseen data interchange and organization stack (cp. Table 7.1) for the *Industry 4.0*.

The architecture of an *OPC UA application*, regardless of the fact, whether it is the server or client, is structured into levels. *OPC UA Security* consists of authentication

Table 7.1 OPC UA versus OSI Stack

Layer	UA	OSI
7	Application layer	Application layer
6	Serialization layer	Presentation layer
5	Secure channel layer	Session layer
4	Transport layer	Transport layer
3		Network layer
2		Data link layer
1		Physical layer

and authorization, encryption and data integrity via signatures. For Web Services the WS-SecureConversation is used and is therefore compatible to .NET and other SOAP implementations. In the binary variant, the algorithms of WS-SecureConversation have been followed and also converted to a binary equivalent, which is termed as OPC UA Secure Conversation. In a mixed implementation, where the code is binary, but the transport layer is SOAP, a compromise between the efficiency of binary code execution and firewall-friendly data transmission is made. The authentication uses *X.509 certificates*, exclusively. The application developer may choose which certificate store the UA application gets bound to. For instance, it is possible to use the *Public Key Infrastructure* (*PKI*) of any active CA directory.

OPC UA bridges the gap between the IP-based world of office automation and the control systems, and devices on the production floor. Interfaces, gateways and the associated loss of information are considered a thing of the past because all production process data are transferred via a single protocol—within a machine, between machines or between a machine and a cloud database. Thus, OPC UA is eliminating the need for traditional factory-level fieldbus systems like the PROFINET.

OPC UA is being further developed as a standard IEC 62541. The *OPC UA protocol specification* (Table 7.2) consists of 15 parts and is still evolving. Therefore at current OPC UA may be considered as a standardization attempt rather than an established standard.

7.5 Goals

Both mentioned standards have successfully been integrated into various business environments. Interfaces to the existing enterprise information processing and automation systems have been built to accommodate the use of novel interoperability standards on all levels (cp. Fig. 7.1). By doing so the improvements in information interchange and advanced reasoning in performing different business scenarios could be achieved, resulting in improved efficiency of the systems involved.

Table 7.2 IEC 62541 structure

Part	Year	Title
1	2016	Overview and concepts
2	2016	Security model
3	2020	Address space model
4	2020	Services
5	2020	Information model
6	2020	Mappings
7	2020	Profiles
8	2020	Data access
9	2020	Alarms and conditions
10	2020	Programs
11	2020	Historical access
12	2020	Discovery and global services
13	2020	Aggregates
14	2020	PubSub
100	2015	Device interface

MES systems massively manipulate data, based on which they strive to optimize production processes. They are being delivered production telemetry data in *real-time* and instantly report production state, produce notifications in cases of malfunctions or stalls, etc. and enable anyone involved into the planning and control of production processes making more informed decisions. Advanced MES systems may offer comparisons with historic data from previously stored production scenarios and foresee upcoming events based on patterns from their knowledge base. To enable this feature, they need to have learning capabilities: dynamic expansion of the knowledge base by relations between procedures, events, and corrective measures on their occurrence. The main goals of implementing MES systems are maximum optimization of the production processes as well as the reduction of unnecessary overheads by vertical integration of enterprise control systems. The main outcomes from their introduction can be summarized by:

- increased planning flexibility,
- stock optimization of raw materials, components, packaging and finished products,
- minimization of incoming materials ullage,
- production process and supply chain management optimization, and
- business process transparency.

Due to the rich service-oriented architecture of OPC UA, vendors and organizations can model their complex process data into an OPC UA namespace. Key industries collaborating with the OPC Foundation currently include pharmaceutical, oil and gas, building automation, industrial robotics, security, manufacturing and process control. The distinguishing characteristics of OPC UA are comprised by being:

- based on a client-server communication,
- focused on communicating with industrial equipment and systems for data collection and control,
- Open - freely available and implementable under GPL 2.0 license,
- Cross-platform—not tied to one operating system or programming language,
- Service-oriented architecture (SOA) compliant,
- inherently complex, since in September 2020, the specification consisted of 3151 pages/15 documents,
- secure, offering security functionality for authentication, authorization, integrity and confidentiality, and
- integrated, by an elaborate information model, representing the foundation of the infrastructure necessary for information integration.

In Industry 4.0 the production digitization process rendered smart machines, production lines and factories producing smart products. Being able to connect to the Internet, they offer better services to their users. On the other side, they also accommodate an *EPC-network* of smart devices, which can be integrated into *Product Life-cycle Management* (*PLM*) as defined by their *Life-Cycle Sustainment Plans* (*LCSP*) in particular and *Integrated Logistics Support Plans* (*ILSP*) in general.

7.6 Synopsis

In this chapter two main objectives were followed. The first was to elaborate the infrastructure of *Industry 4.0*, representing a unified, open as well as cost-efficient alternative to proprietary hierarchical enterprise control systems, and the other to elaborate the *Internet of Things* (*IoT*) infrastructure to support citizens and Industry 4.0 type companies in their collaboration and interchange of ideas and goods.

In addition to the above mentioned information interchange standards, advanced *Peer-to-Peer* (*P2P*) collaboration and message interchange models have been developed (e.g. by the Foundation for Intelligent Physical Agents (FIPA)) for service-based interactions among smart devices. FIPA *Agent Communication Language* (*ACL*) defines autonomous agent's communication and cooperation protocols (cp. FIPA message structure specification) to access and exchange information among IoT devices and enable their collaboration (cp. Sect. 3.3 in Chap. 3). Here, in order to manage information and make it processable by agents, *Web 2.0 ontologies* are introduced to access and organize their knowledge as in the use case described in Chap. 17.

While the IoT is considered crucial for the implementation of the Industry 4.0, there is certainly more to it and the integral solution is envisaged in Chap. 11 on *Intelligent Logistics Systems* (*iLS*).

References

1. Gumzej R (2016) Engineering safe and secure cyber-physical systems: the specification PEARL approach. Studies in computational intelligence. Springer International Publishing. https://books.google.si/books?id=ks1yCwAAQBAJ
2. Drahoš P, Kucera E, Haffner O, Klimo I (2018) Trends in industrial communication and OPC UA. pp. 1–5. https://doi.org/10.1109/CYBERI.2018.8337560

Chapter 8
Logistics 4.0

The digital Internet transformed the way information streams around the world. Internet as a global communication network of networks essentially represents an infrastructure that carries a large variety of services and information resources. Hence, it can be seen as information distribution infrastructure without itself being a service or information source. Only when a plethora of data, services and applications were available on the Internet, it started its exponential growth. After its spread into every aspect of our lives the world slowly started to change from analog to digital.

Logistics, being the invisible backbone of our current way of life, represents the ultimate analog stumbling block on the way to the new industrial revolution—Industry 4.0. Contemporary supply chain management, currently mainly focused on cost reduction, is immensely transforming, due to pressing ecological and social responsibilities as well as the speed-up in technology, new competitors and changes in the world economic pattern. Hence, the need for an Intra- and Interlogistics 4.0 that would be in line with the Industry 4.0.

If Industry 4.0 represents the way of intelligent production with a smart factory as its main building block, Logistics 4.0 represents the way to intelligent logistics with its smart warehouses, transport management systems and their automation as well as integrated supply chain management, as pointed out in the Chap. 6 on e-marketplaces, as its building blocks. They also pave the way to Smart City logistics being elaborated in Chap. 9 on smart cities and communities.

8.1 Introduction

Similar to Industry 4.0, Logistics 4.0 is characterized by increased automation of logistics processes. Starting from warehouse automation and *Warehouse Management Systems* (*WMS*), enabling automated storage units handling, over transport

R. Gumzej, *Intelligent Logistics Systems for Smart Cities and Communities*,
Lecture Notes in Intelligent Transportation and Infrastructure,
https://doi.org/10.1007/978-3-030-81203-4_8

automation and *Transport Management Systems* (*TMS*), to automated *Intermodal Hubs* facilitating multimodal transport.

Moreover, with Kanban-driven *Order Management Systems* (OMS), enabling *Just-in-Time* (*JIT*) replenishment of stocks, advanced *Supply Chain Management* (*SCM*) is becoming more and more transparent, quicker, and easier to perform.

By enterprise information systems integration where *ERP* systems can access data being manipulated by *MES* and *SCM* systems the information flows across companies and their supply chains can be joined. By warehouse and cargo handling automation the physical flows are synchronized with the information flows. The next logical step is to make these systems smarter to be in-line with the trends of smart production, commerce and supply chain management and enable their further evolution by providing for automated vertical and horizontal inter- and intra-enterprise integrations.

8.2 Physical Internet

A radical way to resolve economic, safety, social and environmental goals of contemporary transport is proposed by the concept of the Physical Internet. *Physical Internet* (*PhI* or π) is an open global logistics system founded on physical, digital, and operational interconnectivity, an open shipping system in which modular packages are automatically routed from source to destination through the existing transportation infrastructure [1].

The PhI as a reconfigurable transportation network accommodates roaming intelligent *Intermodal Load Units* (*ILU*), being characterized by their ability to autonomously navigate their way through the PhI. Within the PhI, *PhI Hubs* represent logistic distribution centers with capacities to store and forward ILUs in the form of smart containers [2] (e.g. iTU in Chap. 19).

The idea of the PhI is derived from the foreseen technology transfer—a straight mapping and transformation—of the Internet paradigm onto the logistics and transportation area. Intra- and Interlogistics 4.0 are to be operated by *Cooperative Intelligent Transport Systems* (*C-ITS*) managing ILUs and moving them throughout the PhI in the same way as data packets travel through the Internet. As with the Internet, the main criterion for a successful implementation of the Physical Internet is its *Quality of Service* (*QoS*).

8.3 PhI OSI Model

In order to create a sustainable model of PhI, the infrastructure and protocols of future supply chain operations as well as basic Industry 4.0 and Logistic 4.0 layers need to be orchestrated according to their OSI reference models (cp. Fig. 7.1). In the course of this the *PhI OSI model* features three distinct domains (cp. Fig. 8.1):

Fig. 8.1 PhI domains

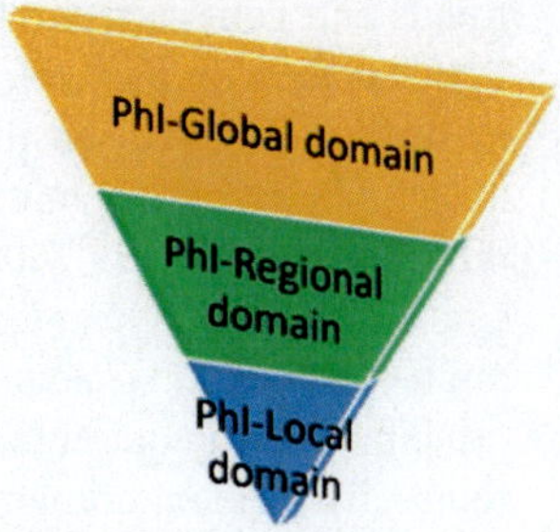

Fig. 8.2 PhI OSI model

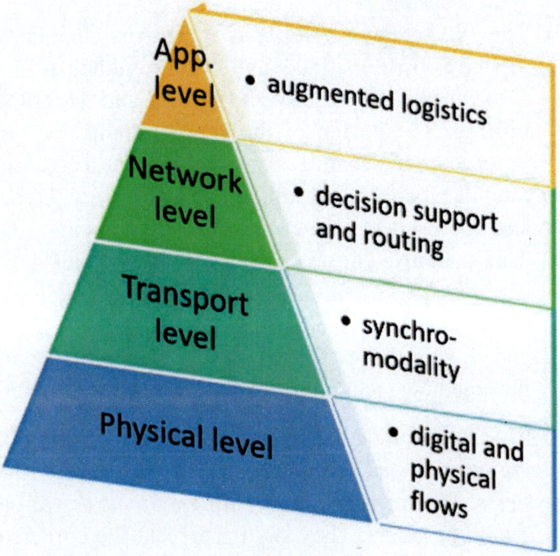

1. global domain (world),
2. regional domain (e.g. European), and
3. local domain (user-defined network),

as well as four basic logistic layers (cp. Fig. 8.2):

1. augmented logistics,
2. decision support and routing,
3. synchromodality, and
4. digital and physical flows.

Even though the PhI is alleged to be the infrastructure and code of behavior for future supply chain operations, the basic logistic layers, need to be defined, including protocols, procedures and best practices of basic future logistic operations:

- The *PhI global domain* is used to define global supply network coordination and collaboration principles. Different scenarios shall be created in order to define trends and general insights. Scenarios amend awareness of change by shedding light on the complex interplay of underlying drivers and critical uncertainties, as well as enhance sensitivity to weak and early signals of significant changes ahead. They facilitate mutual understanding and collaborative action by providing different stakeholders with common languages and concepts in a non-threatening context, thereby opening the space for creating robust, effective and innovative multi-stakeholder strategic operations. The PhI environment will also enclose the principles and requirements to decide between the existing and new transport routes. The global domain shall be used to define and study the criteria that are needed to incorporate economic, social, and ecological constraints into the global PhI ecosystem.

- The *PhI regional domain* is used to strengthen digital and physical flows and to study corridors, hubs and synchromodality. To comprehend the requirements for synchromodal transport big data and geographical forecasting tools shall be utilized. The design of the supply chains for synchromodality is proposed, in order to create seamless and transparent freight transport services, supported by modular loading and unloading technology. Dynamic mapping of requirements shall be used to create more flexible allocation of virtual resources to demand. This shall also consent creation of hybrid channels for parallel distribution.
- The *PhI local domain* is used to deploy *Intelligent Logistics Systems* (*iLS* (c.p. Chap. 11) as the integrating technology featuring augmented logistics as its front end. It shall be used to devise and use new business models that are emerging through new technologies and new supply chain concepts.

*iLS*shall include the application and evaluation of *intelligent agents* and *multi-agent systems, smart devices, Internet of Things, data analytics, big data, autonomous systems, PhI-support and planning systems* as well as *virtualization* and *dematerialization* of assets. They shall represent the cornerstones of *Intelligent Production Systems, Intelligent Transport Systems*, and *Intelligent Governance Systems*.

PhI Hubs shall have a special role in the PhI ecosystem. They represent the main distribution nodes in the whole system, as they shall optimize the alignment of supply chains and synchromodal/multimodal transport. They are to be designed as distributed autonomous systems.

Augmented Logistics (*AL*) shall elaborate mathematical models of multi-objective multi-period intermodal/synchromodal hub operation and location, in particular hub-and-spoke network structures. Such models shall incorporate large scale mathematical programming models for which sophisticated (parallel) solution algorithms are required. Furthermore, augmented logistics shall exploit the mathematical models for robust multi-period service network design taking into account the inputs from the upper level and possible feedback from the lower levels, but will deal explicitly with uncertainties in the demands and in the service supply. The resulting models shall be multi-modal, multi-period stochastic programming models of routing, taking into account transshipment (intermediate facilities) and synchromodality as well as environmental and sustainability measures. The congestion and service levels shall be explicitly integrated within the models.

In order to successfully integrate iLS into the PhI, a sustainable OSI reference model of PhI needs to be created first:

- The *PhI Physical layer* will be responsible for digital and physical flows. Physical flows including modular packaging will be defined by a network of hubs and spokes for all transportation modes. This layer will also allow shippers (manufacturers, retailers), carriers and other providers of logistics services to connect to PhI and optimize the alignment between supply chains and transport services.
- The *PhI Transport layer* will provide hub to hub goods transfer throughout the network by providing services such as flow control, congestion control and reliability of links. It will also be responsible for optimal synchromodal transport.
- The *PhI Network layer* will be responsible for routing control in the network. This layer needs to take into account that carriers have to choose among various routes in their networks, bypassing hubs or taking alternative routes. Therefore, decision support systems will need to allow for optimization of all routing protocols, as well as correct and timely order delivery. Such a *decision support system* would need to be concurrent and distributed, supported by additional AI and machine learning concepts (e.g. intelligent agents and ontologies).
- The *PhI Augmented Logistics layer* will provide for a holistic overview of a truly integrated transport and logistics system through augmented logistics and PhI, as the vision to reach significant advances in terms of efficiency, effectiveness, and sustainability of freight transport and logistics, creating value and adding competitiveness to all manufacturing and retail sectors and supporting the fulfillment of societal goals associated with freight transport.

Special consideration needs to be devoted to data governance, privacy and security, as they will be the most important success factors in the PhI and iLS adoption process.

8.4 C-ITS Versus Augmented Logistics

Augmented Logistics (*AL*) goes beyond the state of the art of *C-ITS* in a number of areas, including:

- Creating a reliable and usable PhI that is fully economically, socially and environmentally optimized;
- Creation of a global environment that will allow for research in PhI and iLS models development;
- Development of capabilities needed to implement the key PhI and iLS concepts;
- Creation of different scenarios for logistic development;
- Implementation of *Open Logistics Interconnection* (*OLI*) model and basic packages that will make them usable;
- Creation of decision support systems that will allow stakeholders to fully utilize their resources in PhI;
- Implementation of data mining and forecasting capabilities into the PhI;

- Creation of new business models to be used with PhI;
- Creation of principles and procedures that will allow for safety and security of freight and associated data within the PhI;
- Integration of smart city, supply chain management, logistics, and transport services;
- Assessment of the impact of new technologies and their integration;
- Optimization of economic, social and environmental constraints in supply chains;

Augmented Logistics will enhance the state of art in smart city, supply chain, logistics, and transport management systems by providing a digital twins environment in which all the scenarios can be tested and verified. *iLS* promotes a novel concept of augmented reality by transformation of logistics resources into services, and by treating them as a commodities.

8.5 Goals

New business models (Fig. 11.2) that shall be created in relation with Intra- and Interlogistics 4.0, or *Logistics 4.0* for short, and are discussed further in related chapters include among others *digital twins*, *supply chain integration*, *production as a service*, *PhI Hubs* taking role in supply chain operations and *postponed manufacturing*.

Augmented Logistics aims to reach for optimal performance of a heterogeneous system composed of several independent and inter-related sub-systems (authorities, transporters and hub operators) at different levels of decision making (strategic, tactical and operational) respecting objectives and constraints of decision makers and stakeholders at each level (social, environmental, economic, viability etc.). It significantly advances the knowledge in area of decision science and operations research and removes barriers towards optimizing systems of independent and autonomous entities with inter-related and inter-twined impacts. Interoperability, data mining, and decentralization of functions will allow augmented logistics results to be used not just in *smart logistics*, *smart production*, and *smart governance*, but also in the emerging *smart cities and communities*.

To grasp these and further issues, *Intelligent Logistics Systems* (*iLS*) should be considered the integrating technology of *Society 5.0*.

8.6 Synopsis

Intra- and Interlogistics 4.0 are currently operated by *Cooperative Intelligent Transport Systems* (*C-ITS*) managing *Inter-modal Load Units* (*ILU*), moving throughout the transportation networks. With the implementation of PhI and iLS, *Intelligent Transport Units* (*iTU*) shall be able to travel throughout the PhI like data packets on the Internet. The PhI as a re-configurable transportation network shall accommodate roaming for iTUs that are characterized by their ability to autonomously navigate their way through the PhI. Within the PhI, *PhI Hubs* represent logistic centers with capacities for the interloading, storage and forwarding of iTUs in the form of smart containers.

As with the Internet, the main criterion for a successful implementation of PhI is its *Quality of Service* (*QoS*). As a consequence, similar QoS indicators used with the Internet and adapted to the PhI traffic can be used as reference. To take this analogy even further, one might conclude that a successful implementation of *Logistics 4.0* would require a *JIT* ordering system with optimization at the customer location.

Eventually, as a modular platform iLS shall allow for further integrations with new research results in related areas, such as smart city, smart supply chain, smart transport and traffic management, etc. In Chap. 19 an example design of such an intelligent transport unit (iTU) is presented from the physical, information as well as procedural perspectives, followed by an evaluation according to the aforementioned QoS principles, especially from the perspectives of efficiency, safety and security. An example of a smart city related smart traffic management solution is given in Chap. 20. An example iLS supply chain management system based on an autonomous multi-agent system platform and QoS as the primary reference is presented in Chaps. 17 and 18.

References

1. Montreuil B (2011) Toward a physical internet: meeting the global logistics sustainability grand challenge. Logist Res 3:71–87. https://doi.org/10.1007/s12159-011-0045-x
2. Montreuil B, Meller RD, Ballot E (2013) Physical internet foundations. Springer Berlin Heidelberg, Berlin, Heidelberg, pp. 151–166. https://doi.org/10.1007/978-3-642-35852-4_10

Chapter 9
Smart Cities and Communities

Since according to [1] most of the world's population is living in cities, we should not be ignorant about their organization as logistic centers of the future *Innovation Society*. It is well known that from the logistics point of view, mega cities are quite a challenge. Let alone the public transport organization (e.g. [2]) that can cope with their inhabitants' needs, other questions remain, putting this modern habitat under a question mark. In the long run such mega cities have proven unsustainable, especially, considering their energy consumption and pollution of their environment. Examples of eco-cities exist (e.g. [3]) where, based on a maximum size of 500.000 residents and resources, scaled to this number, cities have become fully self-sufficient in providing for their own food and energy, offering their inhabitants a healthy habitat with good opportunities for self-realization.

9.1 Introduction

An eco-mega-city should not only provide for a sustainable and ecologically neutral environment to urban dwellers. It should also provide for the necessary infrastructure that would enable them to learn, play and work more efficiently—a *Smart City*. The information infrastructure of a smart city mainly comprises:

- Realistic multimedia content management tools;
- Collaboration tools with intelligent agents and large scale ontologies;
- Fully integrated city-based cloud services and content platforms;
- Smart systems with integrated solutions for education, health, wellness …

The main activities of building a smart green cities from the municipality perspective are currently mainly composed of the following activities:

R. Gumzej, *Intelligent Logistics Systems for Smart Cities and Communities*,
Lecture Notes in Intelligent Transportation and Infrastructure,
https://doi.org/10.1007/978-3-030-81203-4_9

Table 9.1 Key application areas in smart cities (adapted from [4])

City infrastructure and utilities
Broadband DSL, Fiber to the home (FFTH), WiFi
Energy saving (smart grid)
Mobility (smart transport, environment monitoring, real-time alerting, safety)
Innovation Economy
Intelligent city clusters (services, manufacturing, healthcare, tourism)
Intelligent incubators (start-ups)
Intelligent city districts (university campus, techno park, mall, air-port area, city)
Government
Government services to the citizens
Direct democracy (decision making by participation)
Monitoring and measurement (the city as a knowledge base)

- Develop broad-band infrastructure for all public and private facilities located in the city;
- Provide cloud applications for different web intelligence and data processing needs of the citizens;
- Improve infrastructure with smart devices and introduce appliances that allow for information exchange, proximity monitoring, real-time alerts, etc.

What is it about smart cities that makes them so fascinating? The framework of a modern smart city infrastructure comprises several infrastructural layers and application areas (Table 9.1). The infrastructure and utilities of a smart city form the foundations of this modern habitat, which makes it a sustainable place to live, work and play in. On top of this there are innovative solutions for empowering social involvement and innovation, which are properties of the future *Society 5.0*, also known as the *Innovation Society*. While in the *Information Society*, cross-sectional sharing of knowledge and information was still difficult, by the introduction of *data marketplaces* (big data), advanced AI supported *knowledge discovery* techniques, and the use of *extended reality* this should be simplified in the *Innovation Society*.

When building smart cities, there are some possible negative effects, which should be prevented by design. But rather than focusing on them, this chapter is concerned with smart solutions involved and guidelines on how to do things right.

9.2 Smart Homes

The smallest unit of a smart city is a *smart home*. Smart homes integrate a number of smart devices (sensors, controllers, appliances, etc.) which allow for an elevated quality of life and make their maintenance easier.

The corner stone of a smart home application is a *smart controller device*, which integrates all smart devices on the home Intranet. It provides for their universal interface, so they can be monitored and controlled by mobile phones, tablets, computers, or dedicated control devices. The controller device is often realized as a router which maintains a wireless network by which connected devices can access each other. Usually, it is also the gateway, enabling the smart control devices to access and be accessed from the Internet. In order for a *smart device* to become a part of a *smart home system*, it needs to be registered with the controller (e.g. Samsung Smart Home, Domoticz). The controller monitors the state of connected smart devices and coordinates their operation across commands issued by a control device or pre-set operation scenarios. A *smart home application* may be realized as a dedicated back-end application or Web service application with a number of front-end applications comprising Web- and mobile applications, thus building a client-server architecture. For ease of use it may be supplemented by *automatic speech recognition* (e.g. Amazon Alexa) or other *extended reality* devices and applications.

There are a number of benefits from the interconnected devices being able to "talk" to each other. For example, smart lightning may observe the inputs from ambient and external lightning and motion sensors to provide for sufficient illumination of rooms. The sensors may also control the shades to be rolled down during the night. In addition, in combination with location and local weather information, one may set the shades to be lowered at dusk and risen at dawn, and kept shut in case strong wind or heavy rainfall occur. The ambient lightning conditions of different rooms may also vary by brightness and color, e.g. while in the living room the lights may be dimmed and the color set to warm for eye comfort, the working rooms need more brightness and cool colors, especially on work surfaces.

There is more to the interplay between smart devices than just comfort. Imagine the scenario of a wash cycle initiation and leaking water from the washing machine. Without human intervention this scenario might result in considerable damage, whilst in a smart home the smart water sensors should report the leak, and send a signal to the smart controller to stop the washing cycle and contain the damage. A smart home surveillance application may also include an alarm system, which shall detect and report an intrusion as soon as it is detected and possibly provide for video material that can subsequently be used as evidence.

Smart home solutions rely on smart city infrastructure and are interfacing with *smart city solutions*, e.g. water, energy consumption sensors. While various wired and WiFi connections are used to provide the smart home with a broadband Internet connection, the Intranet may be an ecosystem of various wireless networks providing for point-to-point connectivity (e.g. Zigbee, Z-Wave, Bluetooth, ...). The choice of smart device network connectivity currently heavily depends on their producers and their supported smart home solutions.

9.3 Smart Communities

Smart cities, villages, etc. rely on smart city infrastructure to manage their resources more efficiently and provide their inhabitants with a wealth of highly customizable and accessible services. While some are mandatory (electricity, water, ...) the rest are utilitarian or subscription based (cable, wireless communication networks, ...). As with the smart home solutions the various infrastructure networks also suffer from the same problem—proprietary connectivity and data interchange standards. With the unification of smart city data integration standards (e.g. OASC Minimal Interoperability Mechanisms) this situation may improve, but of course it shall take a while, since in the meantime the providers of public services have built their own proprietary networks, and the motivation drivers are weak.

Community services are advertised on the Internet and can be accessed from citizen's homes as well as through various (mobile) platforms. Usually these comprise the services of the local *e-governance* (G2C), public transportation, public health, communal waste and water management, transportation, recreation and sightseeing, etc. In rural areas some community Web services also offer links to e-marketplaces of homegrown food products, which can be ordered online to be delivered to customer location or picked up at producer location by the terms agreed upon online. As a form of direct democracy citizens are offered an opportunity to vote for changes in their local environment by posting their suggestions on-line or via the chosen mobile platform (e.g. SeeClickFix). There are also a number of active aging policies (e.g. assisted living) that can be run through a smart city portal (e.g. WAALTeR).

Besides community Web-pages various platforms have emerged for easier access to personalized community services (e.g. SmartAppCity, FIWOO, etc.), already utilizing the above mentioned new integration standards (e.g. FiWare).

9.4 Smart Mobility

Smart mobility offers smart city dwellers and visitors an insight into the local transportation network to optimize their journeys around the city. The various transportation options offer communities an opportunity to achieve their mobility goals by public awareness and smart traffic infrastructure. For visitors they represent a collection of best transportation options to get them from one touristic site to the other and help them manage their time while in the city.

There are a number of smart sensors and actuators that are involved in smart city traffic management. Smart traffic cams offer the drivers to foresee and avoid traffic jams or avoid accident sites on their route. They may also provide drivers with information about free parking lots. Smart counters assist traffic controllers in predicting and containing traffic jams. Smart traffic lights may especially help pedestrians to avoid longer road-crossing waiting times in low traffic conditions. Bike and car sharing offer mobility by rental and agreed upon return policy. Location based public transportation options offer passengers an insight into the public transportation system to quickly find the next connection to their destination.

Smart mobility should not only consume less energy and produce less pollutants. It shall also be a source of green energy provided electric vehicles could be used as energy store. Moreover, solar-powered e-vehicles are may be considered small power plants. The breaking energy from trains or trams could also be sourced back to the grid, just to name a few examples …

In order to build and maintain the traffic network, smart traffic management solutions should be used that would help forge the most sensible traffic routes and prevent traffic jams while performing maintenance jobs on the network. In order to be able to do so, they should be connected to the smart sensors in place and provide for control in a loop as well as their monitoring and maintenance. Such a smart traffic control solution is presented in Chap. 20.

9.5 Synopsis

There are a number of features that may be offered by a smart city integration platform like the previously mentioned SmartAppCity. To make it transparent, first, a classification of users with their roles in the system is necessary, since each one of them uses the platform in another way:

- City council officials are primarily interested in providing citizens and businesses with information about new opportunities and services. Their duties are mainly concerned with introducing and maintaining infrastructure, fairs, promotions, provided or organized by the city administration.
- Citizens are primarily interested in the elevation of their standard of living in a modern habitat such as a smart city. They are seeking ways to improve their health, get their jobs done in less time and better seize the spare time individually or as part of community services offered by the city or other citizens for the citizens to improve their social lives.
- Businesses are primarily interested in new opportunities arising from some new infrastructure, fairs or collaboration partners within their branch from within or related smart communities.

Last but not least, smart cities can offer an informal way of cooperation between cities, which already established strategic partnerships, and can extend them as smart cities across physical borders of their communities. Smart cities and communities are the foundations of the future *Innovation Society*. They are promoting *crowdsourcing* and *innovation* as the main pillars of the "production" of the future.

References

1. Hassler S (2007) Engineering the megacity. http://spectrum.ieee.org/telecom/security/engineering-the-megacity
2. Guizzo E (2007) How to keep 18 million people moving. http://spectrum.ieee.org/transportation/mass-transit/how-to-keep-18-million-people-moving
3. Cherry S How to build a green city. http://spectrum.ieee.org/energy/environment/how-to-build-a-green-city
4. Schaffers H, Komninos N, Pallot M, Trousse B, Nilsson M, Oliveira A (2011) Smart cities and the future internet: towards cooperation frameworks for open innovation. Springer Berlin Heidelberg, Berlin, Heidelberg, pp 431–446 (2011). http://dx.doi.org/10.1007/978-3-642-20898-0_31

Chapter 10
E-Governance

Creation, distribution, usage and manipulation of information are the major economic, political and cultural activities in the *Information Society*. The development of its enabling information and communication technology and rapid growth of its users results in the exponential growth in the amount of information to be processed, stored and reused. Information dissemination over the internet drastically changed every aspect of social organization, including production, education, health services and (local) government. Because one tends to rely on the enabling technologies and services offered, they also raise multiple safety and security concerns.

10.1 Introduction

The individuals in the *Information Society* are often referred to as *digital citizens* or *e-citizens*, being one of the many trademarks characterizing members of the post-industrial society. The transition from industrial to post-industrial society is characterized by changes in technological, economic, professional, spatial, cultural and various combinations of the aforementioned indicators. Special attention is devoted to the improved service quality, especially the aforementioned safety and security, representing the elevated quality of life.

This chapter is about *e-governance* of the current *Information Society* (*Society 4.0*) as well as new services and governance models for the arising *Innovation Society* (*Society 5.0*). Here, the before mentioned infrastructure in the form of smart devices and services shall be brought together to form a framework of the *Society 5.0* (Fig. 10.1). Next, as an established model of advanced logistics the properties and features of the *Integrated Logistics Support* shall be incorporated to form *Intelligent Logistics Systems* to be introduced in smart environments—*smart production*, *smart logistics* and *smart cities and communities*.

R. Gumzej, *Intelligent Logistics Systems for Smart Cities and Communities*,
Lecture Notes in Intelligent Transportation and Infrastructure,
https://doi.org/10.1007/978-3-030-81203-4_10

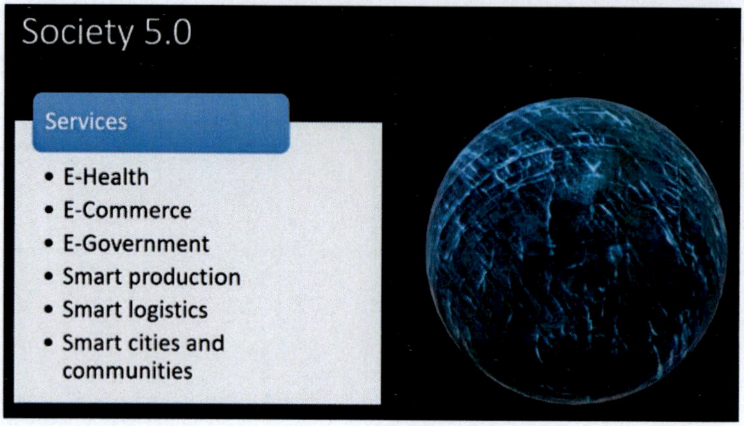

Fig. 10.1 Society 5.0 as a blend of physical and cyber worlds

10.2 Digital Citizens

One of the characteristic forms of workflow organization, often introduced in modern Information Society, is *crowdsourcing*, being a collaboration model that facilitates participatory cooperation culture in municipalities, industry and logistics. As opposed to closed and open source collaboration models it supports and encourages broad participation and cooperation of digital citizens in a global social and economic environment (e.g. SeeClickFix). With its "think globally, act locally" philosophy this model fosters a true innovation society. Its main characteristics are:

- low involvement threshold;
- broad and heterogeneous community;
- volunteering from own interest;
- instantly solving problems being recognized as such by the local or wider society;
- solutions benefits, being recognized and rewarded by the same society.

The success of the contemporary Information Society lies in real-time access to high quality information. Hereby not only one is enabled of making correct decisions, anytime, from anywhere in the world, but also to innovate and create new knowledge. In an innovation society every *e-citizen* is a knowledge worker, contributing to a company's, community's, and global *knowledge base*. In a modern company, innovation is characterized by targeted but extensive information sharing among co-workers. In his/her work process every co-worker receives and shares ideas, being instantly communicated through information systems to other ecosystems of processes—systems of systems [1].

Since the data are more heterogeneous than ever before, a unification technology for their representation, classification, storage and utilization has long been sought. It appears to have been discovered in the form of *taxonomies*, *ontologies* and associated *data marketplaces*. Their management is a complex process which currently still requires human intervention, while their usage can be realized through services of the *Intelligent Web*, *On-line Analytical Processing* (*OLAP*), and *intelligent agents*. Only by this, the stored knowledge can progress in terms of quality.

10.3 Digital Currencies

Money. Everyone knows it and the problems associated with it ...how to earn it, keep it, protect it, ...Being one of the most revolutionary discoveries in human history after wheel invention, it represents the foundation of the world's economy. It is a merit of almost everything that exists in this world. And yet, do we still need it in its current form?

In times of the economic model's progression and evolution money remained the only invariant as a measure of value of work and goods and in its physical form maintained all the strengths and weaknesses it always had. Is there an alternative? In the digital age for sure. Many immaterial financial instruments emerged in *B2B* and *B2C* relations, facilitating the financial flows among companies and their customers as well as among citizens (*C2C*). Electronic currencies slowly but surely enter our daily lives, and are similarly to e-business confronted with "child's diseases".

With the emergence of the *Bitcoin* (Bitcoin) the first *digital currency* was born, enabling individuals and companies to perform financial transactions in a *P2P* manner. Meanwhile there are around 20 active electronic currencies. However, in terms of the monetary system, the digital transformation requires more than just creating a currency, electronic trading accounts, and facilitating money transfers.

In the form of *e-banking e-citizens* would be allowed to access their accounts from their mobile terminals at any time from anywhere in the world. However, this method has a reputation of being prone to hacking, since *electronic wallets* are equally prone to theft and abuse as their physical counterparts. By *biometric identification* it would be relatively simple to ensure the safety of accessing one's account and make credit's transactions on appropriately secured (bank) terminals. From the security point of view, the *e-business* model of performing bank transactions through secure channels seems most appropriate. To ensure secure transactions the "POS to bank" connection safety and security are sufficient. Additional safety can be assured by disabling the insight into the total credit amount through a "POS" terminal. This has already proven successful with credit card payments, where it is only necessary to check the credit state for the current transaction.

Considering the oldest universal means of goods interchange among people, some issues with *e-currencies* still need to be addressed to replace physical money in a digitized innovation society. The biggest of them are organizational, since the world's economy would have to be re-evaluated to form a common currency for which a kind of world bank would vouch for. In terms of *digitization* the worlds monetary interchange system would have to be unified to correctly and securely process all fiscal transactions in real-time. This has already proven to be a major issue with the Bitcoin, since its maintenance requires a great amount of computing resources. In the usual case of multiple e-currencies, one would also have to establish a means of currency exchange, which would again complicate and slow down the monetary exchange processes. A possible solution would be to realize it in a similar form to *High Frequency Trading* (*HFT*) [2], but with additional *security mechanisms* handling remote access to monetary markets (e.g. as in Chap. 19).

Special security requirements pertain to the transaction-acknowledgment system and the credit-account ownership management system. By appropriately replicating the credit-account information in a cloud and inter-checking the transactions between the involved banking information systems, they can be properly addressed. By doing so, all the necessary "infrastructure" for a money-less society would be established. And then, but only then, "money would not matter anymore".

10.4 Digital Economy

Building a national or transnational digital currency should not be too hard, since the value of the world's economy is known. As we have already established above, transferring from the physical form of money to its digital counterpart is only a matter of a political decision of a country, a group of countries, or the world, based on their trust in their own or the common digital economy. Most of the existing mentioned electronic currencies are transnational. Their biggest obstacle is the quantity of electronic money in circulation, since it is currently insufficient, even for a national economy. A denomination in the case of introducing a common world-wide electronic currency would bring serious movements in the allocation of world's wealth, which of course is not in the interest of the world's wealthiest economies. This could however remedy a lot of problems, currently present in economies throughout the world.

Considering the digital economy, in the Chaps. 6–11 we have already covered the necessary information infrastructure to accomplish its goals. As common open information interchange standards are introduced the final analog obstacles are being overcome and the Internet of Things is becoming a reality. The adaptation of old and introduction of new services in the *Industry 4.0*, *Logistics 4.0* and *Smart Cities* will gradually make the systems more manageable and user friendly. To start innovation, appropriate interfaces will have to be established in all the mentioned application areas. Of course appropriate digital laboratories shall have to be implemented to test the new ideas before implementing them in real life. The *digital twins* technology with appropriate *decision support systems* will be crucial for implementing appropriate breeding grounds for new ideas and technologies. Advanced simulation modeling and analysis methods, paired with artificial intelligence methods shall represent the basis of these digital laboratories.

10.5 Digital Government

Apart from advanced *Smart City* services, all other forms of *e-governance* (e.g. e-land register, e-employment office, e-taxes, e-health, e-democracy, …) have already been successfully introduced and can be effectively used by e-citizens. They are granted access based on their electronic certificates that effectively represent them in legal transactions. These are money-less transactions and as such represent no problem at all. As with e-business, the problem lies with transactions involving the concepts of monetary interchange and making money.

Earning money is usually associated with a *debit-credit relationship* with one's employer. Hence, we may say one is *earning credit* as one earns money, representing the reward for an (expected) benefit/work done. A *credit* can also represent a means of exchange in terms that it can be transferred from one person to another in the form of a gift, loan, heritage, etc. But mostly, it currently represents a payment for a work performed by an employee, where the amount is determined by its difficulty and the expected time to finish it. Another perspective of a credit is also being a reward to contract workers, where even for well defined works it is hard to determine their difficulty, whereas on the other hand, due to the specific nature of the employee-employer relationship, it is not possible to set up a monthly allowance. In addition to that, the project co-workers often need to obtain some material or immaterial means to complete their tasks, if their employer does not provide them. By the concept of trust this comes very close to a debit-credit relationship, established when one gets a loan from a bank. A successfully completed task results in a credit in the employee's account, while an unsuccessful completion results in a debt at least as high as a possible advance payment.

When would there be a moment for a new reward model? This is a matter of society maturity. Currently, the modern economy is based on the model of economic growth that is known to be unsustainable in the long run. The proposed debit-credit relationship seems to be a fair but very neo-liberal model of rewarding one's work, since it gives unlimited power to creditors which has already proven negative in the past. An appropriate trade-off between social transfers needed for a human-worthy life in the innovation society and benefits for works and special achievements accomplished, being recognized as such by the community, is sought. According to a universal classification of qualifications and works it would be possible to establish a credit-model of *reward by expected effect* where the effects achieved would enable the community to add some *benefits* to the basic monthly allowance.

Despite digitization there is an interest in keeping and maintaining *property of the earned credits*. As usual, the interest is bigger with the ones, who have more. It is likely that the banks would fulfill their traditional role of safekeeping the credits in the form of electronic currencies on e-citizen's and company's accounts. They shall also likely provide for transfers of credits among different accounts as well as possible conversions between different digital currencies. This would bring the somewhat abstract notion of a "credit" closer to their end users—the digital citizens—and give a meaning to earning them.

10.6 Digital Collaboration

To enable cooperation among partners in the future innovation society a number of information sharing and cooperation platforms have emerged. By the introduction of the *Blockchain* many visibility, traceability as well as security concerns in distributed environments could be efficiently addressed. However, its success and the success of the cooperation platforms, based on it, primarily depends on their broader acceptance and trust placed on them.

There are of course also companies that make decent money out of cooperation platforms. By the rise and fall of the *Bitcoin*, as the first application of the Blockchain, some lessons on the "do"s and "don't"s could have been learned. Hopefully, this will not diminish trust in the technology itself and any of the cooperation platforms built on it. Doing so would set the economy back at least a decade.

The novel collaboration platforms are about to exchange data on a great variety of new services, so one would expect an emerging trend of novel services and platforms. The introduction and breakdown of e-business at the end of the previous century and the rise and fall of the Bitcoin market some twenty years later share great similarity. The time/society was just not ripe for them. Have the society and economy learned their lessons? Time will tell with the successful digitization thereof.

10.7 Synopsis

While in Chap. 9 smart city solutions were discussed with their features to address services to municipalities, citizens and businesses, here, the solutions, being critical for the promotion of innovation, have been presented and discussed.

The interoperability issues of connecting the different areas/subjects of the future innovation society have been highlighted and a solution has been proposed. The solution is composed of using open-source standards for information acquisition, management, exploration and dissemination, combined with the pertaining open/closed source information systems, being used in different environments and imposing the iLS-integration service level above them to provide for their interoperability and sustainability.

From the smart city perspective, the integrated solution should enable the promotion and facilitate roll-out of open-source and crowdsource projects carried out by the community for the community. In this sense the focus is being shifted towards communities rather than cities, since they are the sources and end-users of improvements that they deem necessary to make their lives better.

References

1. Helbing K, Reichel M (1998) Selected aspects of development and planning of production and logistic systems. J Mater Process Technol 76(1–3):233–237
2. Bundesbank D Bedeutung und wirkung des hochfrequenzhandels am deutschen kapitalmarkt. https://www.bundesbank.de/resource/blob/665078/544876d8a09dd548ed15bd74ce14281f/mL/2016-10-hochfrequenzhandel-data.pdf

Chapter 11
Intelligent Logistics Systems

In the past few years, the *Smart City* concept has developed into one of the fastest growing trends worldwide. Researchers from different disciplines are engaged in in-depth analysis and research on the implementation of various intelligent services in an urban environment. As many cities around the world are joining this trend to become a smart city, it is extremely important to examine the changes that this trend is bringing, considering the well-being of residents and visitors. By building a consistent and coherent model of a smart city, one can predict the impact of these changes on urban lifestyle, allowing for better decisions meeting the requirements of the smart city and its inhabitants. Since in the innovation society the boundaries between cities, suburbs and the countryside should be overcome, smart communities are the more sensible term to use. Due to improved logistics and better/new transportation options time/space should be more manageable and the *global village* should become achievable.

11.1 Introduction

As indicated in Chaps. 7–10 on *Industry 4.0*, *Logistics 4.0*, and *smart cities and communities*, *smart services* should be used to implement these concepts as well to achieve interoperability between systems involved in smart production, logistics and city governance. Brought to a common denominator these systems should be represented by *Intelligent Logistics Systems* (*iLS*), implementing the services of intelligent production, transportation and the services sectors.

In addition *Integrated Logistics Support* (*ILS*) should be integrated with the mentioned underlying systems in order to provide for their easier *maintainability* and *flexibility*. In this chapter the structure and functionality of iLS is defined to provide

R. Gumzej, *Intelligent Logistics Systems for Smart Cities and Communities*,
Lecture Notes in Intelligent Transportation and Infrastructure,
https://doi.org/10.1007/978-3-030-81203-4_11

for *interoperability* and *sustainability* of various support systems in *smart environments* of the arising *Society 5.0*, also known as the *Innovation Society*.

11.2 Integrated Logistics Support

Integrated Logistics Support (*ILS*) plans and directs the identification and development of logistics support and system requirements for military systems, with the goal of creating systems that last longer and require less support, thereby reducing costs and increasing return on investments. ILS therefore addresses these aspects of support-ability not only during acquisition, but also throughout the operational life cycle of the system. In terms of longevity and sustainability, this is where military and civil society intersect. The impact of ILS is often measured in terms of metrics such as *reliability*, *availability*, *maintainability* and *test-ability* (RAMT) as well as *system safety* (RAMS) [1].

ILS is a *design methodology* meant for life-sustainment, but it can be much more than that. ILS is the integrated planning and action of a number of disciplines in concert with one another to assure system availability [2]. The planning of each element of ILS is ideally developed in coordination with the system engineering effort and with each other. Trade offs may be required between elements in order to acquire a system that is: *affordable* (lowest life-cycle cost), *operable*, *supportable*, *sustainable*, *transportable*, and *environmentally sound*. In some cases, a deliberate process of *logistics support analysis* shall be used to identify activities and tasks within each logistics support element.

The most widely accepted list of *ILS activities* include:

- Reliability engineering, maintainability engineering and maintenance (preventive, predictive and corrective) planning
- Acquiring resources/Supply (spare part) support (e.g. ASD S2000M specification)
- Support and equipment test/Equipment support
- Manpower and personnel
- Training and training support
- Technical data/Publications
- Computing resources support
- Facilities
- Packaging, handling, storage and transportation (PHS&T)
- Interface design.

Interface design is the relationship of logistics-related design parameters of the system to its projected or actual support resource requirements. These design parameters are expressed in operational terms rather than as inherent values and specifically relate to system requirements and support costs of the system. Programs such as *design for test-ability* and *design for discard* must be considered during system design. The basic requirements that need to be considered as part of interface design include: reliability, maintainability, standardization, interoperability (flexibility), safety, security, usability, privacy, liability (including environmental and hazardous materials regulations).

Integrated Logistics Support (*ILS*) is an integrated and iterative process for developing material acquisition and support strategy that optimizes functional support, leverages existing resources, and guides the system engineering process to quantify and lower life-cycle cost as well as to decrease the logistics footprint (demand for logistics), making the system easier to deploy and support. ILS decisions are documented in a *Life-Cycle Sustainment Plan* (*LCSP*), a *support-ability strategy*, or an *Integrated Logistics Support Plan* (*ILSP*).

ILS planning activities coincide with the development of the system adoption strategy, and appropriate interface design. A properly executed *ILS strategy* will ensure that the requirements for each of the elements of ILS are properly planned, resourced, and implemented [1]. Although originally developed for military purposes, it can also be used for civilian purposes, e.g. in commercial product support or customer service organizations in the form of *life-cycle logistics*. In terms of *smart production, smart logistics and smart governance* systems design and development it shall provide for the necessary organizational and logistics support to ensure that they are properly executed and environmentally consciously run.

11.3 eXtended Reality in Logistics

Smart production, *smart logistics* and *smart* governance represent the *smart cities and communities* infrastructures that should be realized according to the *Integrated Logistics Support* (*ILS*) paradigm utilizing and unifying best practices and sound technologies from each of the environments involved, including sound *Knowledge Management Systems* (*KMS*) and advanced information dissemination techniques.

eXtended Reality (XR) represents a multitude of digital, informational and material services and resources that are complementing physical systems in their native environments. In XR physical and information aspects of operations are treated independently from each other, allowing for the virtualization of direct physical control and the transformation of resources into services.

In XR the location or even ownership of assets or goods are irrelevant. Important is that needed assets or goods are available, when they are asked for. Analogously, in *Augmented Logistics* (*AL*) logistics resources, like vehicles, tracks, roads, terminals, stocks of goods, etc. need to be available, when they are required to perform some logistics operations. Hence, there is a need to separate physical resources from operations or processes, as well as information activities from physical transfers. In plain terms, when a transportation service is needed it will use whatever vehicle is available, whereby the vehicle is treated as a commodity, without regards of its origin, type or ownership. A similar process can be observed with financial services—when withdrawing money from a bank one does not ask for the physically same banknotes that one has deposited to the bank. This not only eliminates numerous pertaining

Fig. 11.1 iLS constraints

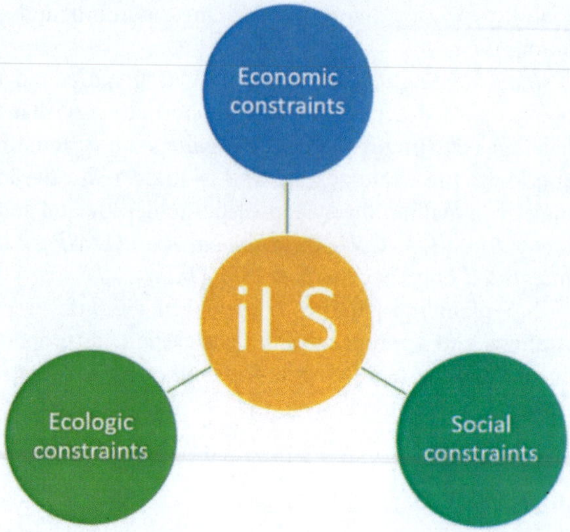

operating restrictions and constraints (cp. Fig. 11.1), but also permits more efficient design of such *Intelligent Logistics Systems* (*iLS*) in smart environments.

The holistic integration and transformation of existing elements and concepts with new services, technologies, business models and standards, is the main challenge of *iLS implementation* (Fig. 11.2). Furthermore, harmonization of logistics systems and logistics resources should be integrated within the iLS to allow for the utilization of shared resources and their abstraction in the form of *AL*.

A truly integrated iLS for sustainable and efficient logistics should be based on the *Physical Internet* (*PhI*) as an open and global system of transport and logistics resources and services operated in an open environment and framework conditions. By incorporating business intelligence on multiple layers of abstraction and application such systems can be considered rational in terms of the economic theory and hence be implemented utilizing smart devices and services and build intelligent agents platforms.

Since the PhI can be regarded as technology transfer of the Internet into the transportation domain, it is possible to make some assumptions that enable comparison between the two systems. Similar to the physical transportation network, the Internet has initially existed as a computer network between heterogeneous computer systems, based on the necessary interconnection and communication standards, to enable reliable data interchange and interoperability. Its rapid expansion has been conditioned by the advent of various applications, like the World Wide Web and E-commerce, that used it a as infrastructure, with additional safety, security and robustness requirements. PhI, although not yet completely established, will develop its full potential in supply chain and logistics domain when its extended reality applications have been devised, enabling its users to maximally utilize its advantages. Eventually,

Fig. 11.2 iLS technologies and business models

its utilization should be measurable by introducing similar service quality assurance mechanisms, as they already exist in the Internet, into the PhI.

AL over the *PhI* consents economies of scale to be carried out through sharing of virtualized resources, by means of flexible allocation of resources and integrated operational management. A complete realization of its concepts means that logistic assets and services cease to be a differentiator, as they are fully standardized, integrated and shared on global level. In other words, supply networks would become a commodity available to any consignor and consignee. In addition, its concepts would add a virtual layer to the various assets based on their types and overlay networks in which they take part, allowing for their full abstraction. In this ultimate scenario, competition would no longer be based on owned and individually-optimized supply chains. Higher-level logistic functions, such as demand-driven network planning, after-sales services, and advanced stock allocation, would drive the competition among supply chain leaders.

11.4 Intelligent Logistics Systems

Generally, the main proposition in production, logistics or governance is that assets, inventory or vehicles for instance, have to exist physically and be distinguishable as individual entities to be used. Eventually, it is important that the assets are available where and when needed, whilst their current physical location and ownership are of lesser importance. By the virtualization of resources and assets one should be able to separate their informational and physical characteristics and consequently separate their availability from their physical or legal form. This would allow for greater flexibility in the utilization of assets as well as system design. Ultimately, by the virtualization of resources new possibilities in the design of smart production, logistics and governance systems would become available together with greater possibilities for improvements in capacity and efficiency.

Simplification and standardization, business practice revisions, and business process harmonization would allow stakeholders in the production as well as transport and logistics domains to integrate their operations and manage their supply networks more effectively, improving asset utilization with lower social and environmental impacts. Business model transformation delivers growth through the power of new technologies. Business model transformation is about doing new things in new ways that create new net-value for employees, customers, business partners and communities.

Massive disruption is on the horizon in transportation and logistics. The dynamics of businesses platforms such as Uber and Lyft, which create value by facilitating exchanges between stakeholders, threaten to massively consolidate the industry profit pool. Similarly, the Apple business model that controls the supply chain of the outsourced Chinese manufacturing is disrupting production as primary business and altering it to a service. New technologies and digitization in transport and logistics will create a plethora of new disruptive business models. Novel enabling technologies, such as the *IoT*, *data analytics*, *big data*, *Artificial Intelligence*, *machine learning* and *Blockchain*, to name only some of them, are being integrated into the business logic. New technological trends shall not only allow for new business models, but will change supply chains as they are today, through dis-intermediation and re-intermediation processes. Intelligent objects will not only communicate with persons, but rather with each other to automate processes by interactions and render fully autonomous behavior.

Smart logistics entities, such as smart vehicles, intermodal containers, cranes and distribution hubs represent the infrastructure of the PhI and are to become the main drivers of Intelligent Logistics Systems. In comparison with digital communications, supply chain and logistics services have a multitude of distinct principal driving factors. Coordination of logistics, transport, infrastructure, and supply networks aim to move, store, supply and use physical objects throughout the world in a manner that is economically, environmentally and socially efficient, secure and sustainable. *Intelligent Logistics Systems (iLS)* are based on physical, digital, and operational interconnectivity, enabled by modularization as well as standardization of interfaces and protocols to provide for interoperability in Society 5.0's smart environments.

11.4.1 iLS Framework

The definition of the overall *iLS framework* encompasses the following phases:

1. Identification of needs, stakeholders and brainstorming to develop a base of stakeholders. For every fraction of the project, the relevant stakeholders (the major players and decision makers) are interrogated and their view on the past, current and future of activities in their geographical zone are acquired.
2. Technological analysis and framework foundation shall carry out an analysis of the existing technologies, their functionality and limitations and the impacts on the efficiency of the current operations. Further analysis of the emerging technologies should also be discussed and systematically included.
3. Functional platform design shall define the platform, the system architecture as well as all technical and functional specifications. Here, the *iLS model* is constructed considering the necessary technical and performance requirements with possibilities for load balancing as well as standards and communication protocols for information interchange among different actors and levels of the project.
4. Simulation modeling and analysis will provide for an environment for conceptualizing and testing the PhI infrastructure and iLS services in a *digital twins* ecosystem. Considering the online nature of most of this data, the use of offline simulation and optimization is somewhat restricted in its ability to uncover real needs in decision-making support. With a combination of simulation, optimization, and data analytics a digital twin—a computerized digital logistics model that represents the network state in any given moment in time, allowing for complete end-to-end visibility—is created. A digital twin represents the nodes of a physical logistics overlay network in cyberspace, based on actual data, and can therefore be used for planning and real-time control decisions.
5. Data governance, privacy and security shall define new forms of supply chain data governance via heuristics and new technologies to ensure the safety of all data as a consequence of the fact that information shall become the most important driver of iLS. Data mining and forecasting will be used to forecast demand for transportation and logistics services. Given the huge volume of various types of data originating from different sources such as industrial case studies, open source data, GIS information etc., this phase is specifically allocated to extract knowledge and visualization from the existing data and assist in learning from it. Input data will be gathered from public and private data and then analyzed using big data tools. Improved security, privacy and trust services will be defined, together with principles of data ownership and management policies. Information and data sharing procedures as well as supportive legal and regulatory practices shall be gathered in the form of implementation guidelines as best practices.

6. Concentration and orchestration shall condition the three previously defined domains: Local, Regional and Global (cp. Fig. 8.1). It is mainly concerned with the concentration and orchestration of PhI concepts on the mentioned levels and their abstractions.

    a. The local domain is meant to create new business models that are emerging through new technologies and new supply chain concepts. Here, key Augmented Logistics (*AL*) objects are investigated—dematerialization of logistics facilities as well as routing in the PhI. Connections to and from PhI nodes from buyers/sellers shall be defined and standardized. This phase shall incorporate autonomous logistic operations from autonomous technologies in warehousing for hub logistics, and then for last mile delivery focused on low-speed operations, followed by autonomous logistics for all modes of self-driving and connected vehicles, trucks, trains, vessels and planes. In addition, automated operations for input and output of intermodal hubs (e.g. billing, tolling, booking, etc.) shall be defined, together with the analysis and processing of IoT and other relevant data from intelligent transport units and autonomous vehicles. A node management system shall be created for smooth, efficient and synchronous flow control in the network.
    b. The regional domain is meant to define all PhI network procedures and conventions. All road, rail, maritime, river and air principal overlay networks need to be included. Predominantly, routing and planning software to accurately route shipments in a dynamic manner across connected intermodal networks shall be created. Algorithms to dynamically match, dynamically altering goals shall be defined, taking economic, environmental and safety requirements into account. Integrative freight network policy shall be defined in order to validate the overlay network models of regional economies and communities. The main outcome shall comprise a basic decision system using financial, market, safety, as well as environmental, ecological and social principles. Here, the criteria for horizontal and vertical supply chain and logistics collaboration shall be formed.
    c. The global domain is meant to define global supply network coordination and collaboration. Characteristics to strengthen digital and physical flows in order to create near-optimal worldwide service solutions shall be investigated. Interoperability and operation standardization will be stipulated. The global implementation of planning processes including demand and capacity forecasting shall be allied with financial, inter-organizational and social considerations. As PhI as a network of networks will change the outlook of global supply chain and transportation networks this domain shall define and scrutinize different scenarios in order to identify trends and general insights. They will be used to investigate current and future transportation corridors. Moreover, analysis of resilience with guidelines on best practices shall complement this domain.

Based on the results from these phases the co-creation and testing environment shall be created in collaboration with decision makers on different levels, stakeholders, end-users, researchers and citizens. The living labs at different partners' premises may allow for exchanges among users, stakeholders, researchers and citizens. User experimentation, evaluation and co-creation shall take place at this stage, while different potential users may experiment and report their feedback thus contributing to future innovative functionalities and services. Finally, the iLS framework shall be subjected to validation and evaluation with the objective to develop a market validation plan, and evaluation reports with ideas for the platform's future development. It will provide an abstraction layer over the PhI to enable its use in the form of extended logistics services.

11.4.2 iLS Integration

The challenge of the *iLS* is to define new services to seize the full potential of the aforementioned smart environments (production, logistics, governance). Global supply network coordination and collaboration is the main goal of the integrated virtualization services. Supply chains are increasingly global and therefore research of logistics in regional and local environment is restricting results and conclusions in relation to the amalgamation of the iLS and *PhI concepts*. Consequently, iLS operate in three different but interrelated domains interacting among themselves and sharing their results.

iLS shall highlight key global trends using demography, economy, resources, environment and technology as global trade drivers. Furthermore, exploration, and transformation of existing and new transport routes in the new *PhI ecosystem* will be the main topic of the global domain. Often the introduction of new services on the global domain may be too complex. Hence, such applications and models shall be created and tested in subordinate domains (local or regional) and then applied to the global framework. Principles of horizontal and vertical collaboration shall be defined and investigated in order to allow all stakeholders to use them in a uniform manner by introducing new scenarios and business models.

Future supply networks necessitate a synchromodal transport system, in which shipments are extensively automatically and optimally routed. The optimization process should take into account many notions simultaneously. Adaptive synchromodal freight strategy on global and regional transport infrastructures shall be defined to assist global supply chains and manufacturing networks as well as final users. Holistic data analytics models shall be created to leverage potentials of data mining and machine learning technologies. Concurrent and distributed systems shall be generated to improve not only forecasting and planning, but to allow for sustainable ways to manage future supply chains.

The basic methodology behind *iLS rationale* is convergent research. The overall *iLS ecosystem* that is crucial to promote and sustain *convergence* draws ideas not only from researchers but increasingly on the cross-fertilization of initiatives between industry, government and general public (using the *Quadruple* and *Quintuple Innovation Helix Framework* [3] of *Innovation Economics* [4]). *Convergence*, in fact, stands for nothing less than the contemporary version of the historical quest for systematic integration of knowledge. iLS shall utilize *Open Innovation 2.0* (*OI2*) [5] as a new paradigm based on a *Quintuple Helix Model* where government, industry, academia and civil society jointly participate to co-create new artifacts together. This model includes user-oriented innovation models to take full advantage of idea's cross-fertilization leading to experimentation and prototyping in real-world settings.

11.4.3 iLS Goals

iLS ambition is to create innovative autonomous systems that will establish transparency in smart environments—production, logistics and communities—to the end user. They shall incorporate *Integrated Logistics Support* to provide for holistic consideration of environmental and economic aspects. By systematic linking of engineering with sustainability assessment and clear representations of the priorities, risks, and trade-offs, iLS shall also enable integration of engineering expertise with *Carbon Footprinting* and *Life Cycle Assessment*.

iLS are a mashup of various technologies: operations research and management, simulation, data analytics, knowledge, machine learning, expert systems, geographic information systems and personal assistant providing simulation and visualization techniques that shall reduce the time from iLS conceptualization to production. The process shall render high-quality *autonomous systems* and *intelligent agents* that can easily integrate and learn from their environment in order to allow their stakeholders to fulfill their goals in a socially and environmentally conscious ways and still remain competitive. Self-healing procedures shall be implemented to allow for quick adaptation in the case of extraordinary events or incidents.

XR shall provide interfaces to *iLS services* that shall allow companies, government, academics and citizens, to easily surpass the challenges that will be imposed by the PhI as a fundamental background process connecting the physical and cyber worlds. Based on expectations made by quadruple helix stakeholders, it shall contain tools that will allow for:

- evaluation of the transport logistic chains and processes and create transparency,
- comparison of planning results and real-time events,
- decision support for an active, target driven and secured action and reaction in real time,
- integration of optimization and simulation models for a scenario-based planning, and
- preparation of plans with tailor-made planning and control functions.

Collaborative modeling shall be used to enable policy researchers, decision makers, industry and academia to address the interdisciplinary convergence approach, and allow for complex systems comprehension, as well as public input in the policy making processes. The creation of an innovation ecosystem will allow participants in a collaborative modeling project develop a deeper level of understanding about the complexity in the policy issues being addressed, increasing their agreement about

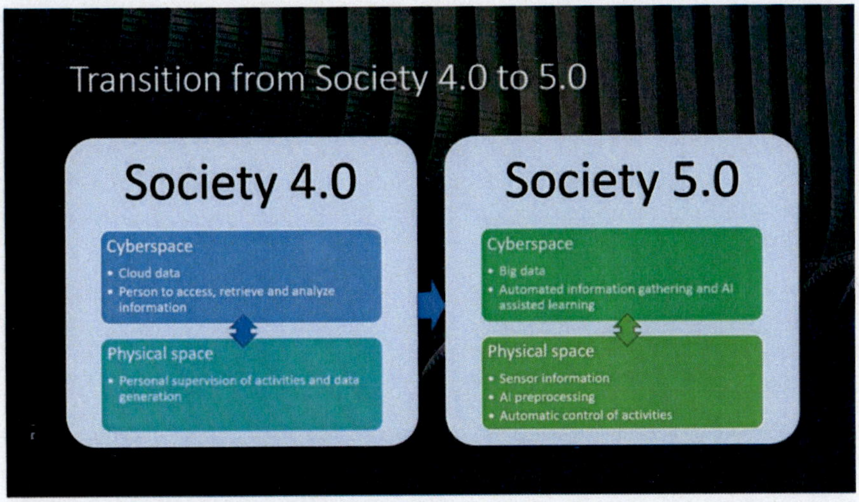

Fig. 11.3 Transition from society 4.0 to society 5.0

root problems and gaining appreciation for the uncertainty, inherent to data and methods for studying complex systems. iLS shall build and maintain a multidisciplinary knowledge base for immediate use by *iLS stakeholders*. This should allow companies to easily prevail over the necessary data analysis skills required to exploit big data which they are currently lacking.

11.5 Synopsis

Smart production (*Industry 4.0*), *smart logistics* (*Logistics 4.0*), and *smart governance* (*smart cities and communities*) are *Intelligent Logistics Systems* (*iLS*) applications used in these smart environments to intelligently manage their processes. They are integrated across the *iLS framework* with the *PhI* into an ecosystem of cooperative intelligent production, transportation and governance systems. *Intelligent agents* and *XR* technologies are to be used to access, use or control them.

Besides smart devices, sensors, actuators and machinery in production environments, and smart logistics facilities, like warehouses, transportation devices, and transport units in logistics environments, intelligent *personal assistant devices*, currently implemented in the form of smart phones or tablets, and accessories like smart glasses or lenses, to support *e-citizens* in their use of *iLS framework*s, are being used. Being spatially and temporally determined, these systems as well as their uses need to be aware of their perimeters for their operation. Only by doing so the signals and messages from their environment and their actions can be adequately assessed and processed.

Besides connectivity and interoperability, these systems need to expose properties, native to *Autonomous Systems* (AS), such as: *self-management*, *self-configuration*, *self-optimization*, *self-protection* and *self-healing*. In order to achieve their goals, they need to maintain *knowledge bases*, defined by *ontologies*, and utilize *Artificial Intelligence* (*AI*) methods for their operation.

These are the building blocks and foundations for the transition from *the digitized Society 4.0* to *the smart and socially aware Society 5.0*, also termed *Innovation Society* (Fig. 11.3).

References

1. ASD/AIA: Sx000i—international guide for the use of the s-series integrated logistics support (ils) specifications. http://www.sx000i.org/
2. Jones JV (2006) Integrated logistics support handbook
3. Peris-Ortiz M, Ferreira JAJ, Farinha L, Fernandes N (2016) Multiple helix ecosystems for sustainable competitiveness: innovation, technology, and knowledge management. Springer International Publishing. https://books.google.si/books?id=N2lBDAAAQBAJ
4. Warsh D (2006) Knowledge and the wealth of nations: a story of economic discovery. W.W. Norton. https://books.google.si/books?id=woPhdVyCArcC
5. Curley M, Salmelin B (2018) Open innovation 2.0: the new mode of digital innovation for prosperity and sustainability. https://doi.org/10.1007/978-3-319-62878-3

Chapter 12
Summary

In the *Innovation Society* the main aim of production will be the *innovation* freeing the modern man of any labor-intensive, repetitive, and time-consuming activities. This is a bold goal, however not unachievable. Many visionaries have already dealt with the idea of smart and green environments in which people could share without feeling restricted or neglected in any way.

Smart habitats have been designed in form of autonomous buildings (e.g. Dymaxion home, etc.) and settlements. Novel forms of transportation have been envisaged to deal with the inefficient use of personal automobiles and poor road infrastructure. Augmented by the Fuller's characterization of the Dymaxion 4D transport: "With such a vehicle at our disposal, human travel, like that of birds, would no longer be confined to airports, roads, and other bureaucratic boundaries, and that autonomous free-thinking human beings could live and prosper wherever they chose", the vision is clear and simple. And he was not the only one …Nikola Tesla has also set the layout of a futuristic green Information Society. He has experimented with wireless transmission of energy (Wardenclyffe Tower) and information (World Wireless System), envisaged anti-gravity means of aerial transportation (Apparatus for Aerial Transportation, Patent no. 1,655,114, 1928), etc.

The idea of low energy, autonomous homes and vehicles is still very much alive today, although almost a century has passed since the first of the aforementioned inventions have been presented. In order to achieve the desired quality of life in green smart cities of the future, besides the infrastructural considerations, the safety and privacy concerns need to be taken seriously. Several use-cases in this book already holistically address and provide solutions to several issues from these application areas.

The infrastructure of the *smart production, logistics and communities* is laid out in this book. The key concepts to doing things smarter than before are represented by the introduction of *Integrated Logistics Support* and *Intelligent Logistics Systems*, which would bring better integration as well as flexibility and maintainability into

R. Gumzej, *Intelligent Logistics Systems for Smart Cities and Communities*,
Lecture Notes in Intelligent Transportation and Infrastructure,
https://doi.org/10.1007/978-3-030-81203-4_12

the aforementioned smart environments. They shall enable the achievement of the much aspired sustainability of the *smart and green habitats* of the future. The results should have a positive impact on social and physical needs of the contemporary society, in which reliable and smart services should represent self-evident means to achieve personal and professional fulfillment throughout people's lives. Freed from the burden of providing for mere living, humans would be able to open their minds and invent freely for the benefit of the society. This is how one could picture oneself the *Innovation Society* of the future: *green, smart, efficient and humane*.

As outlined in the book, there are several critical infrastructures, which are needed to realize such a complex hybrid environment. The most important among them rely on the *PhI OSI model*. Should they be used as a means of targeted attack against any part of these interconnected environments, this would most likely be self-defeating. Any attempts on gaining exclusive control over any of them or misusing any of them could have devastating consequences for the human society. Hence, this should be prevented.

While many topics, addressed by this book, are still very much a work in progress, many have already been implemented for the benefit of digital citizens and communities. We are struggling a bit with some of them, but I would like to pass on a positive perspective on the progress towards the innovation society, since I see no better alternative. We are not there yet, but there is hope, since more and more people are striving for it. Hence, I'm concluding this book as I started it—with confidence that the progress will eventually lead to a better world for our children to live in. Godspeed!

Everything will be OK in the end. If it's not OK, it's not the end.

Chapter 13
Use Case: Augmented Reality

Problem 13.1 How to make efficient use of smart devices and services in *smart production*, *smart logistics* and *smart governance*?

Solution 13.1 By *eXtended Reality* (*XR*) applications, comprising *Augmented Reality* (*AR*), *Virtual Reality* (*VR*), and *Mixed/Merged Reality* (*MR*) applications. When talking about and the join of the virtual and real worlds, we usually mean a visual join (e.g. Fig. 13.1), although contemporary researches also mention the possibility to include other senses like sound and smell as well.

13.1 Introduction

The exemplary AR application in Fig. 13.1 shows an augmented camera view on a mobile phone, where tags of street maps locations are used to highlight landmarks or places of interest. The mobile phone's sensors (camera, GNSS) and actuators (speakers, vibrations, or visual signs) can be used to navigate to a predestined location or simply to explore the surroundings, based on ones preferences. Of course, apart from traffic, pedestrian or tourist maps, data can originate from city infrastructure maps to navigate workers to the nearest hydrant, electrical power transformer, etc. to perform some maintenance works.

R. Gumzej, *Intelligent Logistics Systems for Smart Cities and Communities*,
Lecture Notes in Intelligent Transportation and Infrastructure,
https://doi.org/10.1007/978-3-030-81203-4_13

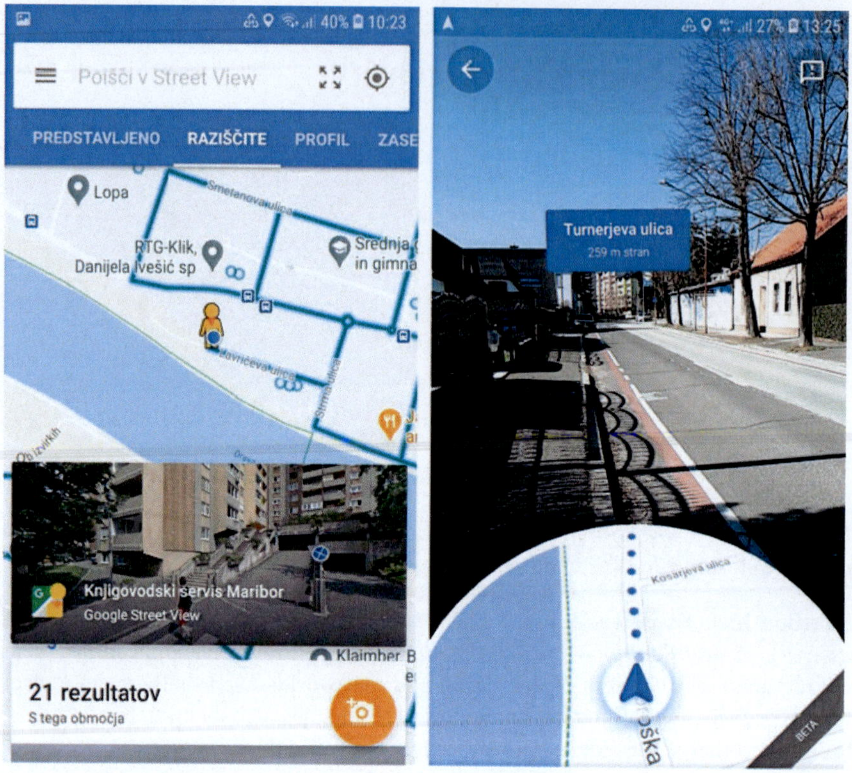

Fig. 13.1 Street view from Google Maps

13.2 Concepts

In [1] the three demands are defined that any *AR application* needs to fulfill:

1. join real and virtual worlds
2. interact in real-time
3. feature three-dimensional representation

The definition may be relaxed in terms of representation, where diverse also non-specialized media may be used (computer monitor, mobile terminal, wearable AR-glasses, etc.). On the other side, it rules out applications that otherwise might seem similar, like watching a soccer game on a mobile terminal with additional information about the game statistics displayed, because most of the information being displayed are not represented in real-time and their representation is not three dimensional. Many motion picture productions which create virtual characters in three dimensional space cannot be considered AR, because they do not fulfill the real-time criterion, etc.

From the theoretical point of view the design of an AR-system is simple—based on user's location and orientation determined by the sensors, the system detects the locations of real objects or markers from the user's perspective; based on these the relevant information on the objects are positioned on the user's visual device using the selected technology. In practice any AR-application comprises three components:

1. *3D world model*: connects real and virtual worlds; it determines, where additional information need to appear in relation to real objects.
2. *Tracking system*: in order to display virtual information with real objects the virtual 3D model needs to be aligned with the real world. This can only be achieved by knowing the exact position and orientation of the user. Since it is usually not fixed, it needs to be monitored in real-time. This is accomplished by a tracking system, realized by different sensors.
3. *Visualization device*: AR information are usually being presented visually on top of the real world; hence, a visualization device is needed, being capable of combining the two views. Depending on the point of occurrence the following classification of devices is being used: on the eye of the user (retinal projection, transparent screen—glasses, lenses), in front of the user (video display on a computer monitor) or on a real object (projection on the object or a specified canvas).

Azuma [1] claims that AR improves the perception of reality, because by AR the user becomes aware of reality-related information that he/she cannot perceive with his/her own senses. These additional information support the AR-user in his/her activities in real-world and real-time. In summary, *AR* technology supplements the human instead of replacing one.

13.3 Applications

Considering contemporary market, different types of *AR devices* are currently being used.

Handheld devices (cp. Fig. 13.1) are increasingly being used in form of smart phones, tablet computers, etc. These devices are equipped with appropriate sensors, compass, GNSS, accelerometer, a camera, and high resolution screen, making them an almost ideal platform for AR applications. Their biggest advantage at the same time represents their greatest drawback—being handheld makes them handy, but not convenient for hands-free use.

Stationary systems (cp. Fig. 13.2) are convenient in situations where bigger screens and resolutions are needed on a stationary location. These systems can be equipped with better cameras, enabling faster and more precise detection of objects. They are also more easily protected against disturbing weather and other environmental conditions. On the other hand, being stationary makes them inappropriate for field work. In Fig. 13.2 Microsoft's Kinect Xbox One technology is being used to precisely

Fig. 13.2 Stationary AR system setup—Xbox technology in an AR fashion industry application [2]

Fig. 13.3 Spatial AR system in automotive industry setup [3]

detect persons in the field of view and use their images as avatars for projecting 3D images of clothes onto them to achieve the virtual dressing room effect.

Spatial AR systems project generated data on real objects themselves. While in their basic setup they are like stationary AR systems, they differ in the presentation of AR data in the real world since by their use any convenient surface, e.g. a table, a wall or even a car (cp. Fig. 13.3) can transcend to an interactive screen. By their development—smaller projectors, better image quality, 3D-projection—new possibilities of combining real and virtual worlds arise. Their biggest advantage is a more precise visualization since the additional information is scaled to the size of the pertaining real-world objects. A great advantage is also to enable multiple users their simultaneous use, enabling real-time cooperation.

Wearable devices are another kind of AR devices, being realized as smart glasses/lenses or watches. As such they can be operated hands-free. However, due to

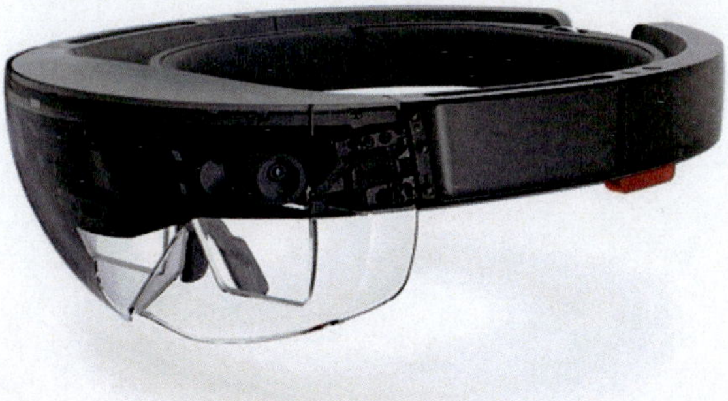

Fig. 13.4 Microsoft Hololens; photo by Ramadhanakbr, distributed under CC BY-SA 4.0 license

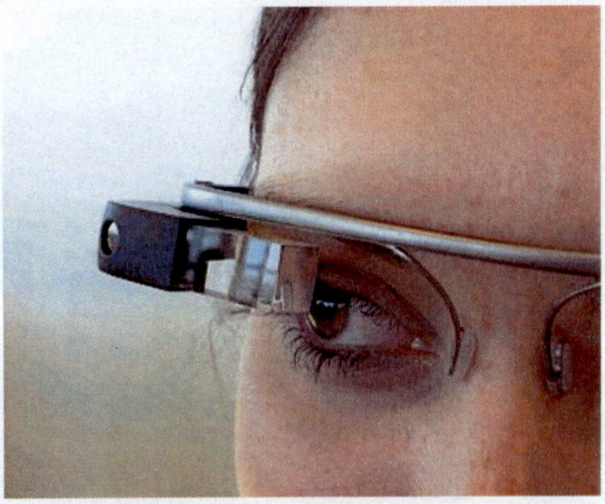

Fig. 13.5 Google Glass; photo by Antonio Zugaldia, distributed under CC BY 2.0 license

their small size they usually lack some sensors and hence rely on a paired handheld device or a back-end system to support their full functionality [4].

Head-up displays are the most broad-spread category of AR devices. They basically consist of a helmet and a micro screen. The device enables its user to simultaneously observe the real world and virtual objects. Contemporary head-up displays enable their users to freely move their head. Multiple installed sensors enable hands-free operation with perfect alignment of virtual objects with real objects. One of the prominent devices in this category is the *Microsoft Hololens*, which came to market in march 2016 (Fig. 13.4). It offers its users the ability to sense arm movement and by this the manipulation of virtual objects.

Smart glasses are similar in construction but smaller than head-up displays. They feature micro screens and additional sensors like microphone, camera, gyroscope, etc. Like head-up displays they enable hands-free operation. *Google Glass* is probably the best known product in this category (Fig. 13.5). Currently it is also the most advanced one since it enables the user to manipulate virtual objects by its three dimensional sensors.

Smart lenses represent the most recent development in AR technology (cp. Fig. 13.6). Major developers in the field like Samsung [5], Google, etc. are already competing in the production of lenses that would project virtual objects into the

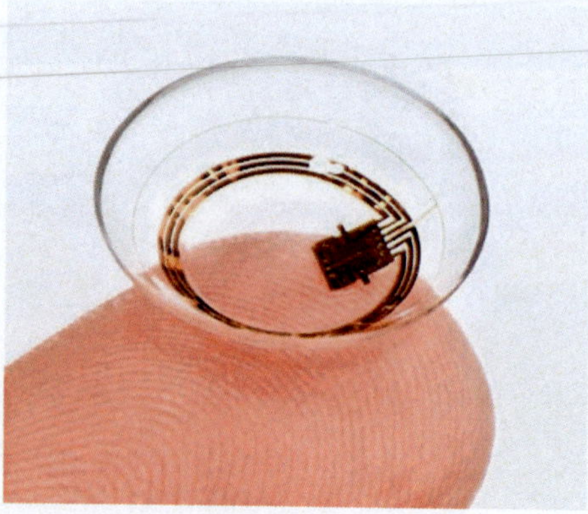

Fig. 13.6 Smart contact lens

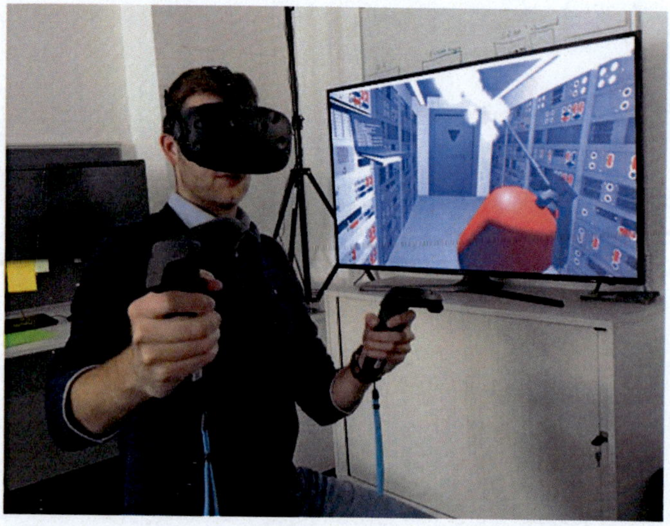

Fig. 13.7 VR-setup; photo by ESA, distributed under CC BY-SA 3.0 IGO license

user's eyesight. The technology seems most compelling in combination with the IoT and 5G communications.

Virtual reality (*VR*) is a simulated experience that can be similar to or completely different from the real world. Applications of virtual reality include entertainment (e.g. video games) and education (e.g. medical or mission training). Other distinct types of VR-style technologies include augmented reality and mixed reality, sometimes collectively referred to as *eXtended Reality* or *XR* [6]. The goal of simulated environments is to prepare the user to function as good as possible in unknown, possibly dangerous situations and environments.

Standard virtual reality systems use either *VR headsets* or multi-projected environments to generate realistic images, sounds and other sensations that simulate a user's physical presence in a virtual environment. A person using virtual reality equipment is able to look around the artificial world, move around in it, and interact with virtual features or items. VR applications usually provide for audio and video feedback, but may also allow for other types of sensory and force feedback through haptic technology. Figure 13.7 outlines mission training in a simulated environment in a VR laboratory setup at the European Space Agency in Darmstadt, Germany.

13.4 Conclusion

Many companies have discovered the advantages offered by AR and VR systems. They are mainly used in user training, but also to minimize human errors in production, warehouse manipulation, etc. operations. The AR based warehouse picking systems have already proven much more reliable and cost efficient that other contemporary warehousing technologies like *pick-by-light* or *pick-by-voice*. From the technical point of view the most important question will be: "how to adapt diverse information systems to work with AR systems?" On the other hand, considering the application areas in the *Society 5.0*, there are no limits and the capabilities of XR are yet to be unleashed.

References

1. Azuma RT (1997) A survey of augmented reality. Presence Teleoper Virtual Environ **6**(4):355–385. https://doi.org/10.1162/pres.1997.6.4.355
2. Trendhunter.com: Augmented reality changerooms. https://www.trendhunter.com/trends/topshop-kinect
3. Easier.com: X-ray vision gives volkswagen a virtual training advantage. https://www.easier.com/81148-x-ray-vision-volkswagen.html
4. Glockner H, Jannek K, Mahn J, Theis B (2014) Augmented reality in logistics. DHL Customer Solutions & Innovation, Troisdorf, Germany, Changing the way we see logistics-a DHL perspective
5. Interllectual.com: Capture the world with smart contact lenses with just a blink of an eye. https://www.interllectual.com/gadgets/capture-the-world-with-smart-contact-lenses-with-just-a-blink-of-an-eye/
6. Wired.com: Get ready to hear a lot more about 'xr'. https://www.wired.com/story/what-is-xr/

Chapter 14
Use Case: RFID Security Stack

Problem 14.1 Can one build a protocol stack that would allow for differentiation of RFID classes according to the applicable security levels?

Solution 14.1 This problem can be addressed using known security mechanisms providing "good enough" security for different purposes, as there are also different classes of RFID tags being used in different applications.

14.1 Introduction

It was soon discovered that, due to moderate storage and processing capabilities, the most broadly used RFID tags are not suitable for security-oriented applications [1, 2]. Does this apply to each and every RFID application or are there differences considering the different classes of RFID devices? Which security protocols can be used with the different RFID classes?

As a possible solution, here the *Public Key Infrastructure* (*PKI*) is used in different usage scenarios, addressing all the different classes of tags individually (c.p. Table 14.1), according to their capability. *Class 0 RFID tags* have been left out on purpose, because their lack of any computing capability makes them unsuitable for security-sensitive applications.

Class 1 RFID tags represent the most common form of RFID devices with a broad application area. Usually, they implement some kind of encoding (e.g. Manchester), for the sake of data transfer safety. These tags are mainly meant for mass identification. Since these tags are read-only, it makes no sense in additionally securing the data on them, however one might wish to make sure, only entitled readers can read their data in order to prevent identity thefts. This can easily be achieved by securing the transmitted data by encrypting them with the receiver's public key. Only certified

R. Gumzej, *Intelligent Logistics Systems for Smart Cities and Communities*,
Lecture Notes in Intelligent Transportation and Infrastructure,
https://doi.org/10.1007/978-3-030-81203-4_14

Table 14.1 Proposed encryption mechanisms by RFID classes

RFID class	Encryption mechanism	Transponder/Reader keys
Class 1	Transponder ID is sent to the reader encrypted with the reader's public key; the reader uses its private key to decrypt the message	ID/Private key
Class 2 (Gen. 2)	Session key is generated and sent to the reader encrypted by its public key; the reader's message is signed by its private key encrypted by the session key	Session key/Private key (signed reader messages)
Class 3	Session key is generated and sent to the reader, encrypted by its public key; transponder's private key is used to sign its message; the reader uses the transponder's public key to decrypt the message	Private key/Public key (signed transponder messages)
Class 4	Session key is generated and sent to the reader encrypted by its public key; the transponder's and reader's private keys are used to sign their messages; session key is used for secure communication	Private key/Private key (signed transponder and reader messages)

receivers would then be able to receive plain text data, while proprietary readers would only get ciphers.

The more advanced *Class 2 RFID tags* offer the possibility to add tracking or handling information to existing transponder data. In order to secure the bi-directional data transfer a lightweight encryption key may be generated in order to secure the transferred data. The reader's public key shall be used to authenticate the reader and authorize it to add data to the transponder's memory.

Class 3 RFID tags are mainly used for admission and process control. Since the transponder needs to be authorized to do something, which is then logged at the reader side, a secure one way authentication and authorization is needed. In order to secure the data-flow and ensure proper authorization, a digital signature is applied to the sent information. To secure the data transfer, the received data is encrypted with the reader's public key. From the reader's side, the transponder's pubic key is sufficient to correctly decipher the received information.

Class 4 RFID tags, which can communicate among each other and smart proxy devices (e.g. smart phones), need strong encryption as well as mutual authentication and authorization. Hence, both ways the transmitted messages need to be properly signed (by private keys) in order to represent meaningful inputs.

Corresponding with the concept of *ubiquitous computing* and the *Internet of Things* novel *Class 5 RFID tags* have been introduced [3], representing active transponders, which may act as readers as well. Since their behavior does not dif-

Table 14.2 Secure identification protocol

Transponder		Reader
Session key generation		(activation of the identification procedure)
Sending the session key to the reader, encrypted by its public key	>	Decryption of the session key with own private key
Sending the transponder ID (and message*) to the reader, encrypted with the session key	>	Decryption of the transponder ID (and message*) with the session key

*In case of Class 1 transponder the message is blank, while with a Class 3 transponder the message is signed with the transponder's private key and needs to be decrypted with its public key by the reader

fer significantly from Class 4 devices, they are considered an upgrade to this class of devices. Their active interrogation with each other requires mutual authentication and authorization. According to the different usage scenarios of these active *RFID transponders* (Table 14.1), two security protocols are presented in the sequel, featuring secure identification and secure communication.

14.2 Secure Identification

In some cases, only one way authentication is needed (e.g. for transponders of Class 1 and 3). Here, the *secure identification* protocol (Table 14.2) can be introduced, providing PGP style confirmation to the receiver. In another words, with the secure identification protocol one can provide for authentication of a transponder with a certified receiver.

Upon getting in the proximity of the reader the transponder generates a session key, encrypts it with the reader's public key and sends it to the reader. Only a certified reader may correctly decypher the session key, which is then used to decrypt the transponder ID and optional message, received with the second transmission. This one way authentication may be supplemented by additional actions on the reader side, based on the message content, being signed by the transponder.

14.3 Secure Communication

In cases, where mutual authentication among transponder and reader is needed (e.g. for transponders of Class 2, 4 and above), the *secure communication* protocol (Table 14.3) can be introduced.

Due to its simplicity, the proposed encryption/decryption function is XOR, since it is symmetrical:

Table 14.3 Secure communication protocol

Transponder		Reader
Session key generation		(activation of the communication procedure)
Sending the session key to the reader, encrypted by its public key	>	Decryption of the session key with own private key
Sending the transponder ID and message* to the reader, encrypted with the session key	>	Decryption of the transponder ID and message* with the session key
Decryption of the reader ID and message* with the session key	<	Sending the reader ID and message* to the transponder, encrypted with the session key
Upon successful mutual authentication continue message interchange; otherwise restart the procedure	–	Upon successful mutual authentication continue message interchange; otherwise restart the procedure

*The messages are signed with transponder's/reader's private keys and need to be decrypted with their respective public keys by the transponder/reader. Subsequent messages only make sense after mutual authentication and authorization of both communicating parties to transfer additional information and perform actions based on them

$$cyphertext = XOR(text, session\,key)$$

$$text = XOR(cyphertext, session\,key)$$

To be consistent with the this protocol is consistent with PGP the public/private keys need to be symmetrical as well. The key length needs to be identical to the length of the transmitted messages in order to enable single cycle encryption/decryption.

14.4 Conclusion

Here, two protocols have been introduced in order to enable strong-enough *RFID security* in two distinct application scenarios. The secure identification protocol ensures that only authorized readers can authenticate transponders and provides resistance to eavesdropping, replay attack, man-in-the-middle attack and even against brutal-force attack, provided the session key length is 96 bits or more. The secure communication protocol provides the same level of security, not limited only to the authentication of the transponder, but moreover for the authentication of the reader, and as a consequence *PGP* protection of communication between them. This is necessary whenever mutual authentication between the communication parties is necessary. This protocol can be applied also to NFC (Near Field Communication) as a derivative of RFID being mostly used for authentication among *Class 5 RFID devices*.

Providing for overall *confidentiality*, *integrity* and *availability* (*CIA*) of information is hard even with enough computing power, but in case of RFID transponders, used in various forms of smart sensors and similar devices, one is limited with their small number of logic gates and amount of power for sourcing them. On the other hand, for every 1000 logic gates, the price of a RFID transponder increases by about $0.05 [4]. Nevertheless, the presented protocols demonstrate that solid security solutions are possible by sensibly distributing and applying appropriate security mechanisms between front and back end systems.

References

1. Peris-Lopez P, Hernandez-Castro JC, Tapiador JM, Ribagorda A (2009) Advances in ultra-lightweight cryptography for low-cost rfid tags: gossamer protocol. In: Chung KI, Sohn K, Yung M (eds) Information security applications, vol 5379. Lecture notes in computer science. Springer, Berlin, pp 56–68
2. McLoone M, Robshaw M (2006) Public key cryptography and rfid tags. In: Abe M (ed) Topics in cryptology - CT-RSA 2007, vol 4377. Lecture notes in computer science. Springer, Berlin, pp 372–384
3. Xiaolin J, Quanyuan F, Taihua F, Quanshui L (2012) Rfid technology and its applications in internet of things (iot). In: Jia X, Feng Q, Fan T, Lei Q, pp. 1282–1285
4. Finkenzeller K, Müller D (2010) RFID handbook: fundamentals and applications in contactless smart cards, radio frequency identification and near-field communication, 3th edn. Wiley, New York

Chapter 15
Use Case: Telemonitoring Vascular Patients

Problem 15.1 Since timely diagnostics and strict follow-up examinations play a key role in proper healing, the introduction of *telemedicine* may be a keystone to a better and more accessible health service.

Solution 15.1 Multidisciplinary cooperation of participating health care professionals, use of common standards in diagnostics and use of clinical pathways in the treatment of vascular patients should provide for a higher quality clinical practice. By the use of telemedicine, a more efficient way of obtaining specialist treatment is achievable. On the other side, its introduction raises security issues, which originate from its enabling information technology. In this use-case a *Cyber Physical System* for telemonitoring vascular patients after applying revascularization procedures in the lower extremities is presented and empirically evaluated. To ensure proper *authentication* and *authorization* of access to diagnostic devices and medical records in telediagnostics of vascular patients an appropriate security protocol has been introduced.

15.1 Introduction

The aging of population, increase of chronic diseases, increasing people's demands for new, more complex diagnostic and therapeutic methods, and lack of health providers are leading the way towards the introduction of new health services, based on new process models and advanced information and telecommunications solutions. *E-health*, *telediagnostics* and *telemedicine* services are promising the better health service for the future, being more effective than existing, established models of *healthcare*.

R. Gumzej, *Intelligent Logistics Systems for Smart Cities and Communities*, Lecture Notes in Intelligent Transportation and Infrastructure,
https://doi.org/10.1007/978-3-030-81203-4_15

By applying the telemedical approach, an efficient way of equally personalized but faster multidisciplinary specialist treatment can be introduced into clinical practice. On the other side, its introduction raises data security issues. Data in the electronic form are easier to access, track and archive [1], but they also travel faster, which can lead to their faster abuse. While ensuring patient security is primarily an organizational concern, in e-health and telemedicine one should also consider information security and making use of the mechanisms provided by the enabling (information) technology. In telemedicine proper information security management should provide for *confidentiality*, *integrity* and *availability* of *medical information* [2].

Provision for security should be an integral part of e-health and telemedicine services, since the healthcare professionals involved are morally, ethically and legally responsible for patients' medical records. In this research a security protocol with biometric authentication and authorization of access to medical devices and data during follow-up telediagnostics of vascular patients has been introduced.

15.2 Methods

Previous telemedical applications in angiology and vascular surgery featured various partial solutions (e.g. one-/two-tire pre-/post-operative medical consultation via e-mail/teleconference [3, 4], three-tire electronic referral, assessment by a practice nurse, and teleconferencing [5]. Our telemonitoring solution features comprehensive telediagnostic procedures, as well as appropriate security protocols with biometric authentication and authorization to access medical devices and data during follow-up procedures of vascular patients.

15.2.1 Telemonitoring Platform

The telemonitoring platform comprises three categories of telemedicine:

1. Wireless acquisition of ABI measurements from a body sensor network.
2. Transfer, storage and processing of patients' diagnostic data and medical records.
3. Interactive telemedicine—real-time video teleconsultation.

The platform (Fig. 15.1) uses *electronic oscillography* to measure fluctuations in blood circulation volume. They can be recorded by infrasonic condenser microphones or electronic receptors. The resulting oscillographic curves are evaluated according to the variations in oscillations on symmetrical parts of the limbs. Important diagnostic evaluation criteria are: the periods, amplitudes and shapes of oscillations [6]. If the difference in amplitudes of the oscillations is greater than 30%, this is an indication of a hemodynamically significant stenosis of the artery. These criteria in combination with visual inspections of the wounds are used at follow-up inspections after vascular interventions.

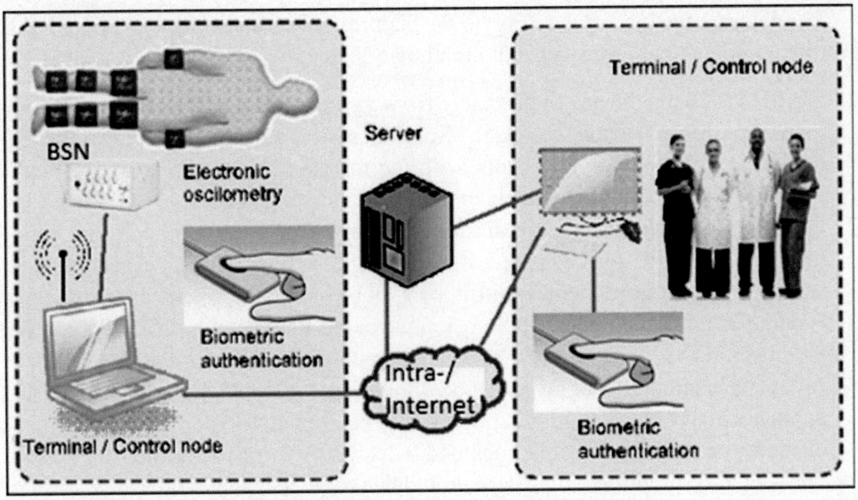

Fig. 15.1 Telemonitoring platform

By electronic transmission of the results of the diagnostic procedure to the medical centers, the effectiveness of medical interventions, being carried out at the secondary healthcare level (e.g. hospital), can be monitored at the primary healthcare level with the aid of a qualified operator (nurse or personal doctor). Specialist teleconsultations with the attending physician (angiologist) by video-conference are possible and hospital visits can be scheduled in case during diagnostics they would prove necessary.

15.2.2 Telediagnostic Procedure

After *authentication* with the diagnostic device, the operator performs the diagnostic procedure on a patient. The results of performing electronic oscillometry on a patient are acquired by a terminal/control node attended by the operator. After the diagnostic procedure is complete, the diagnostic data are transferred to the hospital's medical server, where they are stored to be examined by physicians and hospital staff. The access to patient data is restricted to authorized persons, who are granted access by biometric authentication.

By electronic transmission of the diagnostic procedure's results from the primary to the secondary healthcare level, follow-up monitoring after medical interventions, traditionally being carried out at hospitals, can be performed at the primary healthcare level by a trained nurse. Specialist consultations with the attending physician (angiologist), based on diagnostic data inspection, are possible via videoconferencing. Hospital visits can be scheduled immediately in case during the examination they would prove necessary.

The telemonitoring model (Fig. 15.2) features three main decision nodes that facilitate telemedical follow-up examinations:

1. ↓ ABI ≥ 0.20: on the day of discharge from the hospital, a baseline ABI measurement is performed using the Angio Experience Pro8 device. The patient is scheduled for a follow-up appointment with the telemonitoring office. The patient's medical record is made available to the telemonitoring office.
2. Critical ischemia of the lower extremities: the operator at the telemonitoring office must recognize the five signs of critical ischemia that indicate an urgent vascular surgery: the extremity being painful, pale, pulseless, paraesthetic and paralytic.
3. Healing progression of acute post-operative or chronic wounds: at discharge from the hospital any wounds must be assessed and electronically documented. The following characteristics of a wound should be recorded: location, length and depth in cm, presence of granular tissue, necrosis, exudate, and odor. Photographs of the wounds should also be included. Clear instructions regarding the type of dressing to be used for wound care should be provided by a vascular surgeon. The wound assessment at discharge serves as a baseline for monitoring and treating the wounds at follow-up visits to the telemonitoring office.

Patients with vascular diseases often have associated diseases (diabetes, hypertension, hyperlipidemia), which affect the course of treatment after applying revascularization procedures in the lower extremities. Hence, in a *vascular telemonitoring office*, specialist consultations with different specialists should be enabled via *videoconferencing*.

15.2.3 Telediagnostics Security Protocol

The concept of a wireless juxta-corporal sensor-net on a human body (*Body Sensor Network*, *BSN* [7]), represents the basis of the telemedical platform for telediagnostics of vascular patients (see Fig. 5.1). The network consists of *diagnostic nodes* that gather data on the physical state of patients, being transferred and stored in their medical records on a *Medical Server*, and *control nodes* that access those data, representing data *Terminals* owned by medical doctors, hospital's administrative personnel, insurance company clerks, etc.

To enable secure data interchange among diagnostic/control nodes on one and medical servers on the other side of the network, their mutual *authentication* and *authorization* is necessary. The proposed security protocol reads as follows (cp. Fig. 15.3):

1. Before a node can access any data on the medical server, it needs to be authenticated. In the course of this, it announces its request by sending its entity-identifier to the server.
2. Upon receipt of the authentication request, its plausibility is checked with the server's records and the connection refused in case they are not consistent.

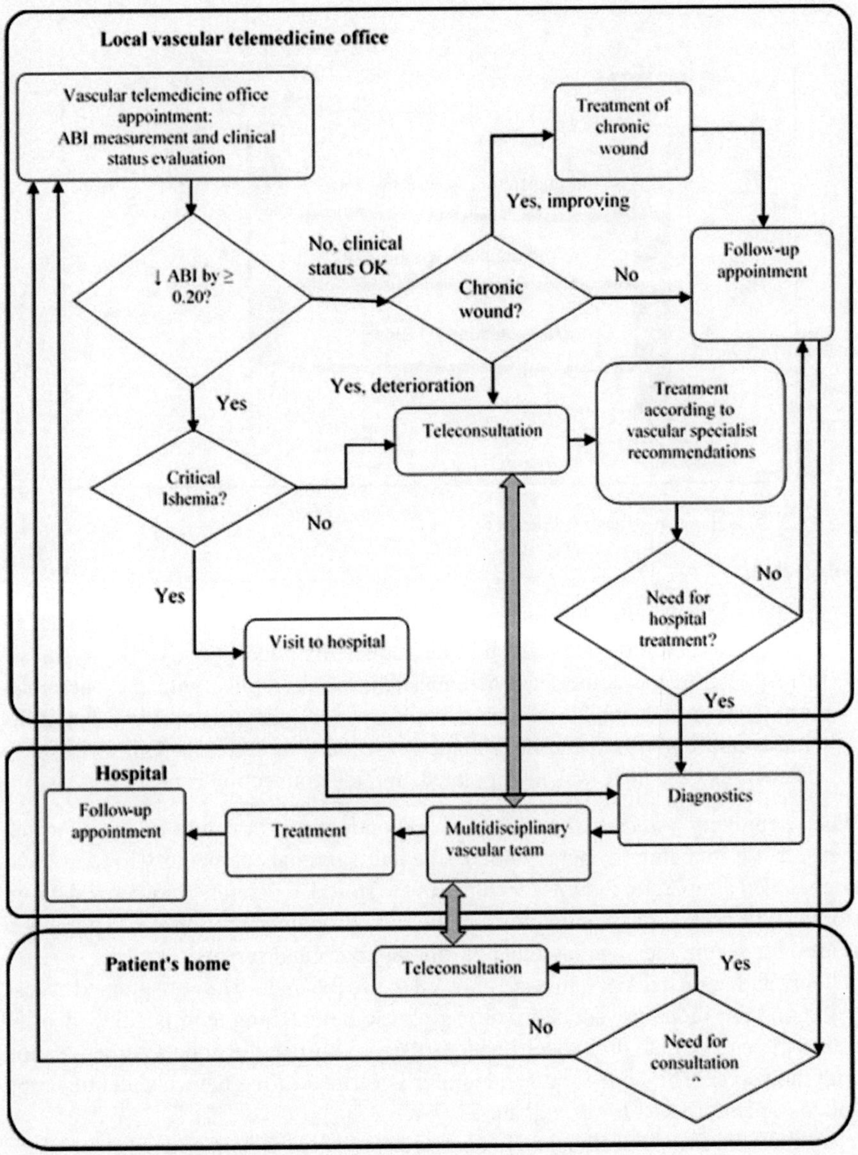

Fig. 15.2 Telediagnostics procedure

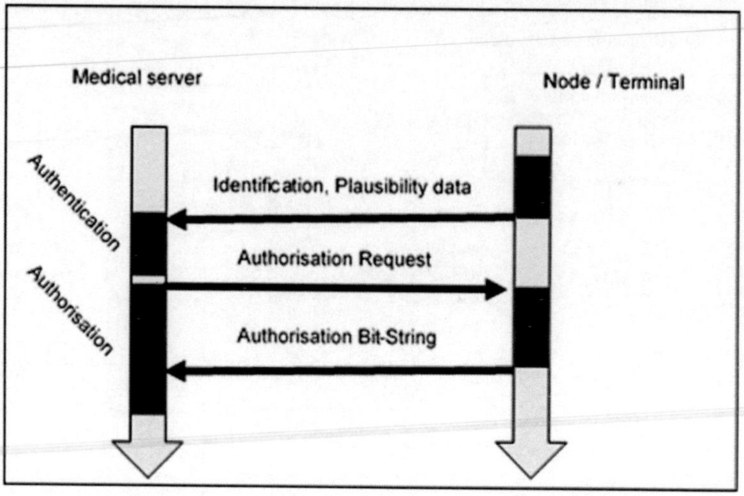

Fig. 15.3 Telediagnostics security protocol

3. In case the authentication of the node was successful, the server generates an authorization request and sends it to the requesting node.
4. In response to the authorization request, the node replies with the operator's authorization data, which the server compares with the authorization data in its records and authorizes the node's access, in case they match. Otherwise, it is assumed that the node was manipulated, and the connection is refused.

When deploying a node, its producer or despondent seals the unit in a way, which prevents it from being tampered with. In the unit's permanent memory its identification, as well as operators' digital certificates are stored. Operators' biometric data are stored at the node for authentication and the server for authorization. *SSL-encryption* is used for secure message interchange among nodes and servers.

Only authenticated operators are allowed to use the node. They are granted access based on their biometric signature or passphrase input. Upon their positive identification the appropriate digital certificate is used to initiate the node's authentication with the server. The node's entity-identifier is formed of the device's identification and the operator's electronic signature.

Authorization, on the other hand, is based on *biometric signatures* of the operators, stored with the server. After receiving appropriate input from the node, the server authorizes the node's access, if they match. Upon successful authentication and authorization, the patient's ID is securely communicated to the server in the form of his/her biometric signature or social security number. As a result of positive patient's identification, the database, containing the patient's medical records is opened for access.

In case the requesting node is a diagnostic device, in the sequel, the database is filled with diagnostic data until the diagnostic procedure is finished and its data stored in the patient's medical record. After that the communication channel is closed. If the communication is interrupted, the connection must be re-established and the data re-sent, following the same protocol.

In case the requesting node is a control terminal, the communication channel is built-up in the same way. Typically, the attending physician or member of hospital administrative personnel uses the terminal to request some diagnostic data from the patient's medical record for evaluation, acknowledges the input of any results from their examination, and closes the connection. In case the communication channel is not closed by the terminal operator, the server needs to close the connection after a certain time interval automatically, in order to prevent an unattended terminal from being misused by an unauthorized party. For any subsequent requests the connection needs to be re-established following the same protocol.

15.3 Discussion

Sharing patients' *medical records* between a hospital on the secondary and a telemonitoring office on the primary healthcare level is essential for effective telemonitoring of vascular patients. The telemonitoring office should operate daily, following an established schedule with instant reporting of diagnostic results to the secondary level. Vascular surgeons should inspect the incoming diagnostic data at least twice a day to efficiently monitor the healing process of the patients to determine their further treatment as well as the time intervals between follow-up examinations.

Mutual multifactor authentication and authorization has proven effective in protecting stored diagnostic data to be shared between the telemonitoring office and the hospital information system. Since biometric characteristics are unchangeable and non-detachable, it is impossible to lose and to misuse them. Hence, biometric authentication and authorization of nurses in the telemonitoring office and vascular surgeons at their hospital department is considered both secure and effective. The doctors on call should be able to examine the diagnostic results from the telemonitoring office twice a day and offer teleconsultations during their shifts.

The Angio Experience Pro 8 diagnostic application program could be extended according to the telediagnostics procedure (Fig. 15.2) to automatically notify the operator in case of an indication for intervention. The healing progression of chronic wounds can be effectively monitored by a nurse, provided that detailed records and photographs of the wounds are electronically accessible and clear instructions about appropriate wound dressings are provided by the vascular surgeon.

In terms of efficiency, the the study has demonstrated that the telemonitoring office can take over most of the follow-up workload. The vascular surgeons are disburdened to be able to focus on new patients' inspections and performing procedures. Since the schedule for surgeries then only depends on the availability of operating teams, rooms and free beds at the vascular department, it is easier to set-up. Consequently, the patient waiting times at the vascular department can be significantly reduced.

15.4 Conclusion

By the proposed approach, follow-up examinations, currently being carried out by *vascular surgeons* at the *secondary healthcare level*, can be transferred to the *primary healthcare level*, and be performed by *trained nurses* using *telemedical equipment*. The main benefits of this approach are comprised by the facts that patients can be treated in their local environment, the waiting times for initial visits at the vascular surgery and angiology clinic can be reduced, and, as a consequence, more time can be allotted to patients undergoing vascular procedures. By this approach also the cost of their medical treatment can be reduced by minimizing the number of follow-up appointments with vascular surgeons.

Health information in electronic form is easier to access, transfer, track, store, process and protect. However, it can become a matter of misuse without proper protection. To some degree ensuring information safety and security in e-healthcare and telemedicine is an organizational concern. In addition, to provide for confidentiality, integrity and availability of patients' medical records (*EMR*), the proposed security protocol should be used. Telediagnostics procedures lead to medical findings being stored in patient's medical records. Secure electronic medical records including specific data regarding medical treatments and diagnostic findings are of crucial importance to medical professionals, patients and society.

Biometric authentication is considered a contemporary method of protecting medical data stored in hospital information systems. Since biometric characteristics are unchangeable and non-detachable, it is impossible to lose them and misuse them as such. Hence, biometric authentication of persons in *public services* (*e-governance*), especially in *healthcare services*, is considered both secure and effective. In combination with electronic diagnostic devices it shall enhance their availability as well as traceability of changes in medical records.

By secure telemedicine and telediagnostics applications the patients would gain trust in personalized *e-health* applications and the notion of better healthcare service due to shorter waiting times and reduced need to travel to obtain specialist treatment. By the presented approach to telemonitoring of vascular patients, the transfer of their follow-up treatment to the primary level of health care was enabled in a safe and secure manner with benefits for the patients, physicians and society.

References

1. Harrington L, Kennerly D, Johnson C (2011) Safety issues related to the electronic medical record (emr): Synthesis of the literature from the last decade, 2000—2009. J Healthcare Manag 56(1):31–44. https://doi.org/10.1097/00115514-201101000-00006
2. Narayana Samy G, Ahmad R, Ismail Z (2010) Security threats categories in healthcare information systems. Health Inf J 16:201–9. https://doi.org/10.1177/1460458210377468
3. Schmidt CA, Schmidt-Weitmann SH, Lachat ML, Brockes CM (2014) Teleconsultation in vascular surgery: a 13 year single centre experience. J Telemed Telecare 20(1):24–28 PMID: 24352901

4. Palombo D, Mugnai D, Mambrini S, Robaldo A, Rousas N, Mazzei R, Bianca P, Spinella G (2009) Role of interactive home telemedicine for early and protected discharge 1 day after carotid endarterectomy 23(1):76–80. https://doi.org/10.1016/j.avsg.2008.06.013
5. Hands LJ, Clarke M, Mahaffey W, Francis H, Jones RW (2006) An e-health approach to managing vascular surgical patients. Telemed e-Health 12(6):672–680 PMID: 17250489
6. Kröger K, Gröchenig E (eds) Nicht invasive Diagnostik angiologischer Krankheitsbilder, 3rd edn. ABW Wissenschaftsverlagsgesellschaft
7. Miao F, Bao S, Li Y (2012) New trends and developments in biometrics: physiological signal based biometrics for securing body sensor network. In Tech

Chapter 16
Use Case: Supply Chain Operating Network

Problem 16.1 How to achieve global visibility and manage a company's supply chain on the Internet through a central repository?

Solution 16.1 Through a Web-portal or a cloud solution, implementing a *Supply Chain Operating Network* (*SCON*) using *Platform-as-a-Service* (*PaaS*) and *Software-as-a-Servic*e (*SaaS*) services, respectively.

16.1 Introduction

According to [1] automation of supply chain management systems in the past few years has been a major concern of both—academia and industry. With increasing competition in the global market, organizations have realized that it is not enough to merely improve their efficiency internally, but it is also necessary to consider the competitiveness and efficiency of their supply chains. Performance measurement and optimization play a vital role in improving the competitiveness of supply chains [2]. Moreover, with the rapidly developing world economy and global marketplaces, a drastic increase in the pressure on organizations to find new ways to create and deliver value to customers through supply chain management has been detected. The recognition of the importance of building relationships with customers for improvements in profitability, as well as serviceability and reduced costs throughout the supply chain has grown [3].

Different authors have [4] emphasized the importance of delivering superior total value to the customer in terms of promptness, cost, quality and flexibility rather than by focusing solely on promptness and cost. In relation to this service quality has become a major attraction for both practitioners and researchers [5]. Its proven relationship with improving business performance by lowering costs, increasing cus-

R. Gumzej, *Intelligent Logistics Systems for Smart Cities and Communities*,
Lecture Notes in Intelligent Transportation and Infrastructure,
https://doi.org/10.1007/978-3-030-81203-4_16

tomer satisfaction, achieving customer loyalty and increasing profitability has further motivated them to explore this area.

The biggest problem that modern organizations face today, is their inability to quickly adapt their supply chains to changes dictated by fast-evolving markets. More specifically, organizations currently rely on various mechanisms to perform quality assessment of their partners and adapt their decision-making processes to improve their business performance. Hence, numerous research studies have dealt with the subject of service quality in supply chains and a consensus has been reached concerning a strong relationship between service quality and supply chain performance [6].

To achieve the desired global optimization throughout the entire supply chain, one must employ proper mechanisms to measure, monitor and control service quality through an inter-organizational assessment system [7]. Such systems are sensibly implemented to identify opportunities for improved supply chain efficiency and competitiveness, to help understand how companies operating in supply chains affect each other's performance, to support the supply chain in satisfying consumer requirements and to assess the results of an implemented initiative. As collaborative networks, supply chains also have the need for decentralized operations where partners conduct their planning autonomously and only exchange a limited amount of information. A frequently proposed solution for this purpose are auction mechanisms, which are most effectively realized by the concept of multi-agent systems.

16.2 Elemica E-Marketplace

Elemica is a good example of a *SCON*, developed to help the companies to achieve the best control of their global supply chains through information sharing between the business partners in order to create value across the supply chain. In 2014 Elemica processed a turnover of about 230 billion $ annually of more than 6500 trading partners. Elemica started in 1999 with only 22 companies, but it became a large network integrating leading chemical companies in the global market. It provides solutions to integrate the supply chains, starting from e-procurement, vendor inventory management, order processing, and collaborated forecasting. As indicated in [8], the level of integration may vary from link to link within a supply chain. Hence, Elemica provides for different levels of integration:

1. *Quick link* network:
    a. *Quick link ERP* to connect the ERP system of a member to the ERP systems of other connected business partners.
    b. *Quick link E-mail* allowing the business documents to be transferred automatically between various members through Elemica. This represents a self generating documentation solution, which is preventing any possibility of human errors while creating the documents and moreover, allowing for more time for order processing. The documents can also be signed automatically—if required—depending on local legislation.

c. *Quick link portal*, being a web-based solution enabling the members to interact online and allowing for easy access from anywhere through the Internet. The online portal can prevent a member from placing an order above the production capacity of the supplier, so it is a powerful interactive tool for achieving the best operation control system.

2. *Smart link* applications respond to the specific needs of the supply chains in terms of:

   a. *Customer management*. In small organizations, the order processing can be done manually, while in large organizations this is not possible. Here, due the large number of orders and ordering volumes, any human error while processing the orders can be very costly. The ideal practice would be to utilize an automated order processing tool. This application helps large companies perform *sales orders management*, including vendor managed inventory (VMI), repeat orders, Kanban systems, and delivery schedules.
   b. *Logistics Management Solutions*. As the logistics information are dynamic in nature, they are changing over time. Hence, it is crucial for any supply chain to be constantly updated with all the changes of these information. This tool of Elemica enables all participants in a supply chain to receive continuous updates on their logistics operations. This updates are generated automatically in electronic form, hence, eliminating the paper work and any manual activities to get the needed logistics information such as phone calls, e-mails, faxes, etc. This application contains all the features of *logistics and transport management*: road transport, ocean transport, time slot booking, warehouses, carrier rates, and freight invoices.
   c. *Supplier Management*. Like the Customer management tool manages supply chains downstream, supplier management helps manage supply chain upstream connections. This application enables automated *purchase order management*: purchase orders, vendor managed inventory, demand forecasting, global supply, and commercial invoices.
   d. *Sourcing Management Solutions*. Sustaining a large network of companies, Elemica provides a very effective and powerful tool for sourcing. This application may be engaged with the purchasing activities throughout the supply chain to provide for the most effective approach to *materials sourcing*: Self Service Bid, Supported Service Bid, Accelerated Sourcing, Assisted Strategic Sourcing, Strategic Sourcing, Collaborative Sourcing, Special Projects, and Logistics Sourcing.

In summary, the quick link is suitable for small scale supply chains, while the smart link is designed to handle large scale supply chains.

16.3 Discussion

According to our case study of Reda chemicals [9], the benefits of their operations after B2B integration through Elemica were:

- Increased *agility* of the supply chain of Reda through *timely actions* and *prompt responses* to the customers' needs.
- Achieved high levels of *visibility* through information sharing with all the business partners.
- Reduction of time consuming manual activities through automated *logistics and transport management.*
- Efficient operation handling through a high level of *flexibility*.
- Reduced overall inventory and minimization of the bullwhip effect, using *vendor managed inventory*.
- Reduced human errors across the supply chain by *automated order management* activities.
- Enhanced collaborative planning with the business partners through *correct, real-time* data sharing.
- Improved *materials sourcing* activities within a larger supply network of potential suppliers and/or customers.

By the Elemica's B2B automation solution, information are collected at each echelon of the supply chain and are available in real-time. Since late data are mostly wrong, their dismissal was a major improvement in Reda's *order fulfillment cycle*.

Considering business *data security and confidentiality*, often being discussed as an issue in *B2B collaboration platforms*, in Elemica not all data are presented to all users. Only relevant data that are to crucial for their decisions and for performing their transactions are visible to registered users.

16.4 Conclusion

With respect to the mentioned benefits, one may consider SCON a considerable upgrade to the previously existing EDI based B2B solutions. Unbound from proprietary ERP solutions, they represent a successful supply chain management solution for *vertical e-marketplace integrations*, where members of the same industry can share their products and information, and collaborate in the local, regional as well as global markets to accomplish their goals.

References

1. Pal K, Karakostas B (2014) A multi agent-based service framework for supply chain management. Procedia Comput Sci 32:53–60. https://doi.org/10.1016/j.procs.2014.05.397. http://www.sciencedirect.com/science/article/pii/S1877050914005973. The 5th international conference on ambient systems, networks and technologies (ANT-2014), the 4th international conference on sustainable energy information technology (SEIT-2014)
2. Reimann M, Zhang W (2013) Towards a multi-objective performance assessment and optimization model of a two-echelon supply chain using scor metrics. Cent Europ J Oper Res 22. https://doi.org/10.1007/s10100-013-0294-7
3. Niraj R, Gupta M, Narasimhan C (2001) Customer profitability in a supply chain. J Market 65(3):1–16. http://www.jstor.org/stable/3203463
4. Ketchen D, Rebarick W, Hult GTM, Meyer D (2008) Best value supply chains: a key competitive weapon for the 21st century. Bus Horiz 51:235–243. https://doi.org/10.1016/j.bushor.2008.01.012
5. Deshmukh SG, Seth N, Vrat P (2006) A framework for measurement of quality of service in supply chain. Supply Chain Manag 11:82–94. https://doi.org/10.1108/13598540610642501
6. Chow G, Heaver T, Henriksson L (1994) Logistics performance: definition and measurement. 24:17–28
7. Seth N, Deshmukh SG, Vrat P (2006) A conceptual model for quality of service in the supply chain 36. https://doi.org/10.1108/09600030610684971
8. Lambert DM, Enz MG (2017) Issues in supply chain management: progress and potential. Ind Market Manag 62:1–16. https://doi.org/10.1016/j.indmarman.2016.12.002. http://www.sciencedirect.com/science/article/pii/S001985011630308X
9. Rashad W, Gumzej R (2014) The information technology in supply chain integration: case study of reda chemicals with elemica. Int J Sup Chain Mgt 3(1):62–69

Chapter 17
Use Case: Autonomous Supply Chain Management System

Problem 17.1 While one may consider SCON a good choice for vertical e-marketplace integrations, horizontal e-marketplace integration is still not as efficient. Is there a solution that would enhance their efficiency and would apply equally well to both horizontal and vertical e-marketplaces?

Solution 17.1 This use case proposes mediation among autonomous agents representing nodes in a supply network based on the *SaaS* principle as a more natural approach to supply chain management. By this approach nodes emerge on the network, notify their environment of their abilities, establish vertical and horizontal connections to other nodes, and perform supply chain activities, as long as they exist.

17.1 Introduction

Autonomous control is seen as an option to handle the increasing complexity and dynamics of supply chains [1]. In [2] *autonomous control* is defined as a "processes of decentralized decision-making in hierarchical structures". It presumes interacting elements in non-deterministic systems, which possess the capability and possibility of rendering decisions independently. The objective of autonomous control is the achievement of increased robustness and positive emergence of the total system due to distributed and flexible management of dynamics and complexity. Hence, autonomous control increases the flexibility and robustness of the *Supply Chain Management* processes by enabling *decentralized decision-making* and supply chain operations execution at the nodes of a *supply network*. Therefore, *Supply Chain Management* platform middleware needs to provide for the necessary plug-and-play functionality to seamlessly interact with the relevant modules of *enterprise*

R. Gumzej, *Intelligent Logistics Systems for Smart Cities and Communities*,
Lecture Notes in Intelligent Transportation and Infrastructure,
https://doi.org/10.1007/978-3-030-81203-4_17

control systems, thus allowing for *Augmented Logistics* in the form of autonomous *Supply Chain Management*.

By the use of open *electronic data interchange* (*EDI*) interfaces (e.g. JSON) and *intelligent agent frameworks* (e.g. JADE) the nodes can easily communicate information and cooperate among each other. Through these open interfaces the nodes can roll out transactions from their underlying *Enterprise Resource Planning* (ERP) systems, enabling organizations to seamlessly collaborate and contribute their products/services to the global *e-marketplace*.

The goal of this research was to investigate a possible improvement of overall supply chain service quality by introducing an autonomous multi-agent supply chain management platform, supplemented by an appropriate *decision support system*. The agents would utilize multiple-criteria decision support systems to evaluate the *quality of service* (*QoS*) of their partners maintained throughout the history of their transactions. Initially, the decision model parameters are balanced. Since depending on supply chain strategies QoS criteria may be conflicting, the Analytic Hierarchy Process (AHP) [3] is used, allowing each echelon to fine-tune its QoS-evaluation model according to its own preferences, while at the same time maintaining the consistency of the decision model.

By doing so companies would be enabled to improve their supply chains by assessing the service quality of their partners. Since more informed decisions are taking place continuously and autonomously within the entire supply network, service quality is being improved throughout entire supply chains up- and downstream, automatically (cp. Chap. 18).

17.2 Autonomous Supply Network

According to [4], supply chain configuration is inherently complex due to the dynamics, decentralized and distributed characteristics of supply networks. Each node in a supply network possesses several alternative collaboration options, consisting of autonomous organizations with unique resources, capabilities, objectives, and competencies. To enable autonomous control in cooperation between "digital companies", here, a new concept of an *autonomous supply chain node* is introduced (Fig. 17.1). According to their roles (e.g. raw materials provider, components provider, producer, distributor, retailer, etc.) their control needs to encompass the specific sets of logistic skills, knowledge and behavior—the logistics competencies of different types of echelons within a supply chain.

There are two kinds of properties pertaining to autonomous supply chain node modeling—static and dynamic. While their static properties describe their structure, the dynamic properties represent their behavior in the supply network model.

The static node properties outline the composition of the relevant supply chain object and are usually the starting point to its modeling. They are usually outlined by *UML class diagrams*, where classes of objects with their properties and operations are

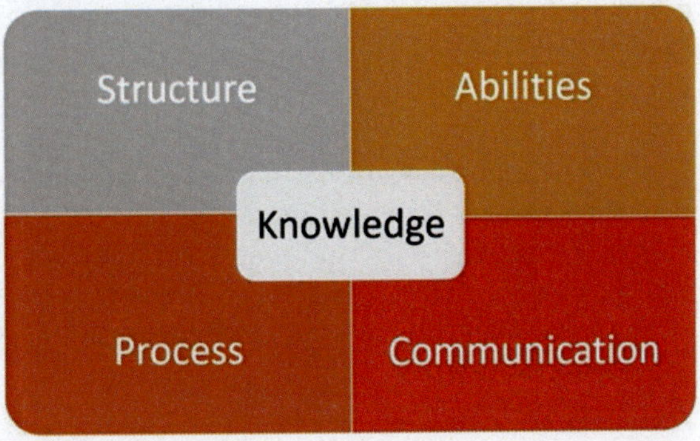

Fig. 17.1 Autonomous supply chain Node's logical model

defined. They enable three basic types of abstraction—classification, generalization and aggregation.

The dynamic node properties outline different skills, knowledge and behavior that are necessary to produce the supply chain outcome (cp. Fig. 17.1):

- *Knowledge* represents all knowledge, which the supply chain objects need to possess to allow for decentralized decision-making. It should combine spatial and temporal aspects. For simple knowledge representation UML class diagrams [5] are sufficient, while for more complex and dynamic connections a dedicated knowledge representation language (e.g., *OWL*) needs to be used.
- *Abilities* represent the abilities of the individual supply chain node. Supply chain management operations represent an important part of supply chain node's abilities. These abilities need to be regarded as abstractions of problem types and scenarios, occurring in reality, and represent the business logic of the nodes.
- *Processes* represent the logic-temporal sequence of activities and states of supply chain nodes. Here, the node's processes are modeled [6]. They play a central role, connecting the static and dynamic node's properties, specifying the behavior of supply chain nodes. According to [7], the appropriate notations to model processes are *UML activity diagrams* as well as *UML state diagrams*. In supply chain management a reference process model has been defined—*Supply Chain Operations Reference* (*SCOR*). It is based on five distinct processes: *plan, source, make, deliver* and *return*.
- *Communications* represent the contents and temporal sequence of information exchange messages between supply chain nodes [8]. Modeling their communication is especially useful to represent all possible interactions of these otherwise loosely coupled systems. To model their communication, *UML sequence diagrams*, showing the progression of message interchange protocols in time, as well as *UML class diagrams*, representing message contents, may be used.

Autonomous supply chain nodes may be considered *autonomous systems*, integrated into an overlay supply network (cp. Chap. 3). The autonomous manager's functions are intertwined with the functions of the *SCOR model*, since for managing the autonomous elements its functionality is analogous to managing the production/servicing processes [9]. Hence, the autonomous manager's functions *monitor, analyze* and *plan* collectively represent the SCOR activity *plan*. While performing this activity, they also use and provide information to the autonomous node's knowledge base. In response to the inputs and in correspondence with its knowledge, the autonomous manager's *execute* function calls up autonomous element's operations to perform the SCOR activities: *source, make, deliver* or *return*. By associating supply chain networks with overlay networks and their nodes with autonomous systems, we end up with another implementation of autonomous systems in the field of supply chain management. In the sequel this use-case is briefly elaborated and evaluated.

17.3 Decentralized Agent-Based E-Marketplace Platform

Agents are considered one of the most important paradigms that on one hand may improve current methods for conceptualizing, designing and implementing software systems and on the other may represent a solution to the legacy software integration and automation problems. *Multi-agent architectures* are fundamental mechanisms, underlying the autonomous components that support effective behavior in real-world, dynamic and open environments [10]. In artificial intelligence, an *intelligent agent* is an autonomous entity which observes through sensors and acts upon an environment using actuators (i.e. it is an agent) and directs its activity towards achieving goals (i.e. it is *rational*, as defined in economics [11]. Intelligent agents may also learn or use knowledge to achieve their goals. By their behavior they are classified as *model-based learning agents*.

In this research a *decentralized agent-based e-marketplace platform* [12] has been devised based on the JADE (Java Agent DEvelopment) framework [13]. In contrast to centralized e-marketplace platforms, the agent-based approach is distributed and dynamic, and hence corresponds more closely with the supply chain's natural behavior. In addition, by utilizing service quality-oriented *decision support system* and making increasingly informed decisions, the service quality within the entire supply network of contributing companies is expected to gradually improve. The two autonomous systems properties involved in this process are *self-optimization* and *self-healing*.

17.3.1 Collaboration Scenario

As companies arise, they are assigned agents that represent them in the e-marketplace. They are registered with a yellow-page service corresponding with their line of business in general and their products/services in particular. The *yellow-page service*, being used by all company's agents on the e-marketplace, manages the routing of their requests for quotations. As companies close, they are unregistered from the yellow page service.

Once the quotations from supplier companies agents have arrived, a selection and ordering procedures are conducted between a customer and its supplier agents. By a shipment information a purchasing transaction is completed. The goal of the selection procedure is to improve overall supply chain service quality by allowing companies agents to evaluate the service quality of their partners through the history of their transactions. Any transaction in progress is automatically subjected to a service quality-oriented decision support tool, where the current transaction data, supplemented with historic data on the service quality of prospective suppliers, are evaluated with the goal of best meeting customer's needs. Consequently, since more informed decisions are taking place continuously and autonomously, service quality is being improved along the company's supply chain up- and downstream.

17.3.2 Service Quality Assessment

Although there are already numerous and well established models in the literature, measuring supply chain performance is still subject to investigation. While most models consider pure performance in terms of costs, a moderate number of scientific works are considering the *quality of service* (*QoS*), being mostly referred to as *total quality*. When considering agent-based platforms, most researches focus on the QoS of agents themselves. As opposed to the established client-server model of e-commerce, the delegation of services to agents is known to balance the computational load, otherwise heavily affecting the performance of the established client-service architectures.

This research relies on the holistic approach to service quality assessment as defined in [9, 14]. The main reasons for employing a holistic approach lie in consequence of the otherwise identified absence of connection with a strategy, lack of system thinking, in which a supply chain must be viewed as a whole entity, lack of balanced approach integrating financial and non-financial measures, as well as loss of supply chain context, thus encouraging local optimization. Based on the decentralized agent-based e-marketplace platform for inter-organizational (supply chain) cooperation and the service quality assessment model, supply chain management operations may be performed autonomously, according to the aforementioned goals.

Autonomous service quality assessment requires measurable performance indicators, pertaining to the service quality realized among supply chain partners (nodes) of a supply chain. Since some of the performance indicators defined in [14] are qualitative, they are not suitable for automated service quality evaluation. Hence, only the quantitative service quality indicators form the basis of the QoS evaluation. As a measure of QoS, q has been introduced as a 4-tuple of service quality indicators (C, T, D, P) representing service quality on every link between a chosen pair of supply chain nodes. All mentioned service quality indicators were assigned an equation that best expresses their interrelations with the node's inputs (e.g. burden/load) as well as with other service quality criteria. Finally, the overall service quality indicator (Q) has been expressed as a normalized sum of these criteria.

Correctness represents the proportion of correctly completed transactions:

$$C = c/d \tag{17.1}$$

where c represents the number of correctly fulfilled orders and d the total number of orders. Timeliness represents the proportion of timely completed transactions:

$$T = t/d \tag{17.2}$$

where t represents the number of on-time fulfilled orders and d the total number of orders. Timely fulfilled orders are those whose processing time (tout-tin) is less than the time specified for delivery. Dependability represents the sum of availability (A) and reliability (R), defined in the sequel:

$$D = (A + R)/2 \tag{17.3}$$

Availability represents the proportion of time a node is in a functioning condition (available to its customer's orders):

$$A = 1 - MTTR/MTTF \tag{17.4}$$

where $MTTF$ represents its average Mean-Time-To-Failure and $MTTR$ its average Mean-Time-to Repair (e.g. maintenance operations or other exceptional situations, which prevent a node from responding immediately to an incoming request). Reliability can be expressed as the opposite of a node's fault rate:

$$R = 1 - F \tag{17.5}$$

where F represents the ratio of failures to fulfill requests (e.g. due to out-of-stock situations). Reliability is considered the ratio/percentage of cases when the observed process in a supply chain did not fail to complete.

The price of an item (P) should be divided by the highest offered price ($max P$) to render the price indicator for our service quality evaluation (Q_p), namely:

$$Q_p = 1 - P/max P \tag{17.6}$$

Finally, the overall service quality can be calculated as a normalized sum of these indicators:

$$Q = (C + T + D + Q_p)/4 \tag{17.7}$$

Since the individual summands in Eq. (17.7) are all in the (0.1) range, in order to normalize the result, it is divided by 4. While fine-tuning the model, we might decide to ponder the individual summands, as indicated in [14]:

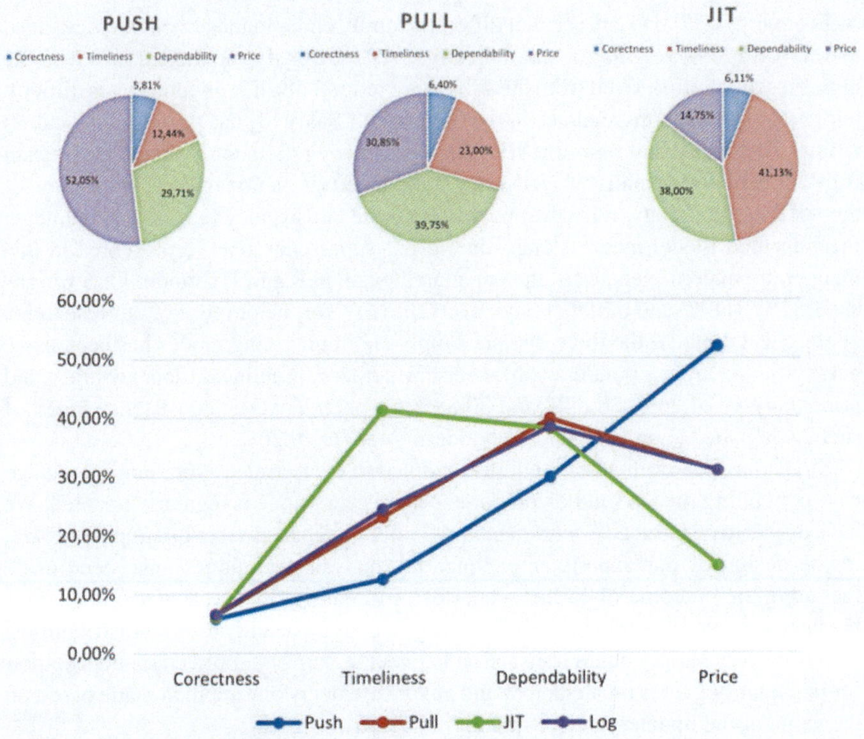

Fig. 17.2 Weights of indicators

$$Q = (cC + tT + dD + pQ_p)/(c + t + d + p) \quad (17.8)$$

As there are at least three different production modes (push, pull and just in time (JIT)) in supply chain management, weights of the quality indicators naturally differ (cp. Fig. 17.2). For example, in push replenishment mode price is most important, in pull replenishment mode dependability is most important, while in JIT replenishment mode timeliness is most important.

A meaningful pondering may be different from application to application, however considering supply chain management operations, a sensible option is consulting some experts and building an *Analytic Hierarchy Process* (*AHP*) model [15] that represents the best consistent blend of their knowledge. AHP compares every two indicators at a time, instead of comparing all indicators simultaneously. All indicators are pairwise assessed in order to derive their relative weights. In this context relative means that indicator weights are calculated in relation to each other. In our case for each method, five experts have done the evaluation. Comparisons have been arranged in a matrix and then the final matrix has been created using geometrical mean calculation.

From Fig. 17.2, one can see that different supply chain management methods lead to very different weights for the indicators. As previously mentioned, because the numeric values are derived from the subjective judgments of individuals, it is difficult to prevent some inconsistencies in the final result. Saaty [3] has proposed the calculation of a consistency ratio. In AHP the consistency ratio should not be larger than 0.1. In practice, especially when dealing with experts with no previous knowledge of the AHP methodology, it is customary to treat the consistency ratio as a percentage, meaning that model inconsistency should not supersede 10%. Represented in this manner, the inconsistencies of the calculated push, pull and JIT-modes QoS weighs are 1.87%, 1.49% and 0.88%, respectively. Finally, for the purpose of our research a geometrical mean of the three distinct supply chain operating modes has been used, having the following ponders values for correctness, timeliness, dependability and price: 6.49%, 24.26%, 38.37% and 30.88%, respectively. The consistency index of such a "logistic" (Log in Fig. 17.2) pondering renders 0.26%.

Since our QoS evaluation model is involved in every transaction, our knowledge base containing the QoS indicators for each and every node is regularly updated. We have used it to construct a *Bayesian model* for our QoS based *decision support system*. In the design of our experiment, sequential analysis techniques have been used, including the outcome of earlier scenarios in the design of the next scenario. This has been achieved by updating beliefs throughout the prior and posterior distributions.

After every supply chain transaction between a pair of supply chain nodes, their service quality gain is recalculated, and any future decisions are then made based on the value of the updated service quality indicator:

$$QG_{i,j} = QG_{i,j-1} + Q/n_i \tag{17.9}$$

where i represents the corresponding supplier, QG its service quality and j the instance when the current transaction Q value is added to the corresponding QG, whereby n_i represents the total number of transactions of the corresponding supplier node.

17.3.3 Knowledge Sharing

Ontologies were developed in artificial intelligence (AI) to facilitate knowledge sharing and reuse [16]. They enable a shared and common understanding of a defined domain that can be communicated between people and application systems. In our case, it is meant for conceptualizing and managing supply chain knowledge, agent communication protocols by which this knowledge is shared between supply chain nodes and autonomous agent behavior models.

Our ontology establishes the basic vocabulary of terms, which are relevant in *e-commerce*; e.g.: company, item, inquiry, quotation, order, etc. The concepts it introduces are associated with supply chain nodes, represented by agents and transactions that are exchanged among them. Ultimately, there are only two fundamentally distinct

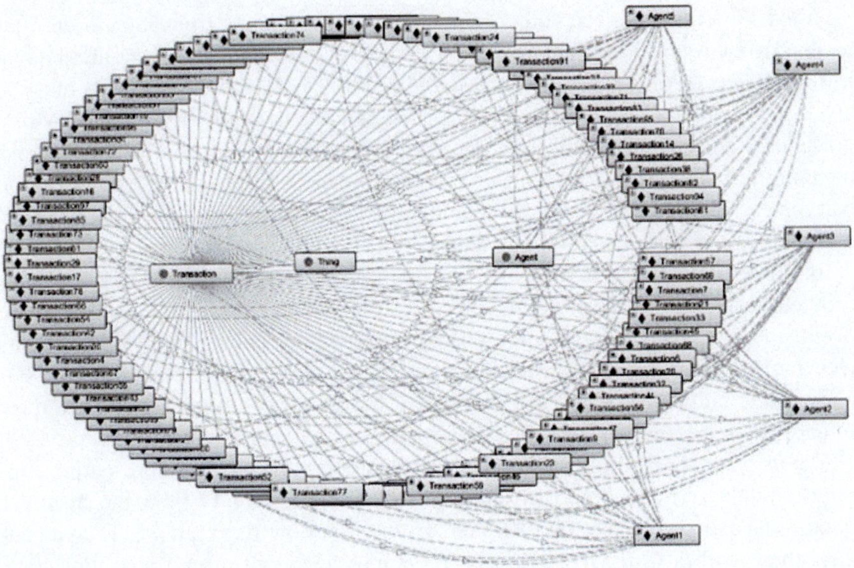

Fig. 17.3 E-marketplace ontology [12]

classes of objects in our ontology—*Transaction* and *Agent*. Hence, any individual object in our *e-commerce ontology* cannot be an instance of more than one of these two classes.

A key component of any multi-agent based system is communication. Each multi-agent system needs a mechanism for knowledge sharing. Our *e-commerce ontology* may be used to store this knowledge in a cloud and make it accessible to services through agents. Agent communication protocols may be utilized to access this knowledge and to establish sensible relations among supply chain nodes as well as to implement appropriate behavioral models.

In our experiment (cp. Fig. 17.3), we have defined 100 transactions and split transaction objects evenly among our 5 supplier objects (Agents 1 to 5), being members of our customer object's (Agent) supply chain. In the sequel, agent communication protocols that are needed to perform supply chain operations and utilize knowledge shared by the e-marketplace ontology are presented.

17.3.4 Communication

Agents need to be able to communicate with users, system resources and with each other to collaborate and negotiate [10]. According to the *Foundation for Intelligent, Physical Agents* (*FIPA*) [17] each agent has its own characterization, which determines its behavior.

Agents interact with each other by messages, using *Agent Communication Language* (*ACL*). Agents' messages represent *actions* or *communicative acts*, also known in *speech acts theory* as *performatives*. To be fully FIPA compliant, agents must be able to receive any legal FIPA-ACL message and at the very least respond with a not-understood message, in case the processing of the message does not render a meaningful result [16]. FIPA-ACL messages contain a set of one or more message parameters. Which and how many parameters are involved in a meaningful agent communication varies according to situation.

In our experiment, autonomous agents communicated over the HTTP transport protocol using the *FIPA-contract-net Protocol*, which according to [18] allows one agent, the *Initiator*, to have some tasks performed by one or more other agents (*Participants*), and further optimization of the function that characterizes the task. For a given task, any number of Participants may respond with a proposal; the rest must refuse. The Initiator then continues negotiations with the participants that proposed.

Let us consider the indicated collaboration among five agents, representing suppliers, and the one agent representing a customer (cp. Fig. 17.4). First, the customer obtains the data on the potential suppliers from the yellow-page service. Then it contacts them with a *Call for Proposal*s (*CFP* message) indicating transaction data. The supplier objects reply with their proposals (PROPOSE message) according to the transaction data. After comparing the incoming proposals and selecting the most appropriate one, the customer object informs the sender that it accepts its proposal (ACCEPT-PROPOSAL) and the others that it rejects their proposals (REJECT-PROPOSAL). The selected supplier fulfills the order of the customer by informing it (INFORM message) about the fact that appropriate fulfillment actions have been initiated.

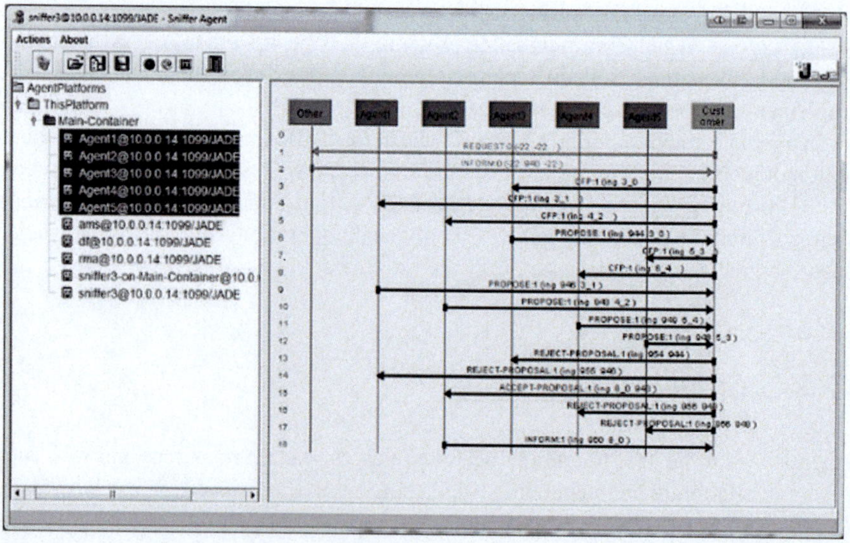

Fig. 17.4 Procurement action as performed by the customer agent from its 5 supplier agents [12]

In any order-cycle the customer and supplier nodes take over the Initiator and Participant roles in the autonomous agents' communication, respectively. Transaction data is transferred between them with appropriately tagged messages. The service quality for any supply chain node is maintained by its agent within the knowledge base. Since initially there is no previous service quality data, the decisions are made solely based on the offered price. With any subsequent order being fulfilled, however, the service quality is recalculated and the proposal selection decisions are made in an increasingly informed manner.

17.3.5 *Behavior*

All Agents are generic until they are enlisted as members of a supply chain (Fig. 17.5). When agents are registered in the yellow page service, their specific role (raw materials supplier, components supplier, distributor, producer, consumer) is indicated. A partnership between a pair of supply chain nodes can be observed as a *customer-supplier relation*. For example, a customer orders an *item* (a material, a service or a finished product) from its suppliers and decides upon their offers based on the respective price and service quality. *Transactions* represent business transactions among the individual nodes of the supply chain. These transactions are represented by messages exchanged between *Agents*, representing supply chain nodes, and may involve goods/services or monetary transfers.

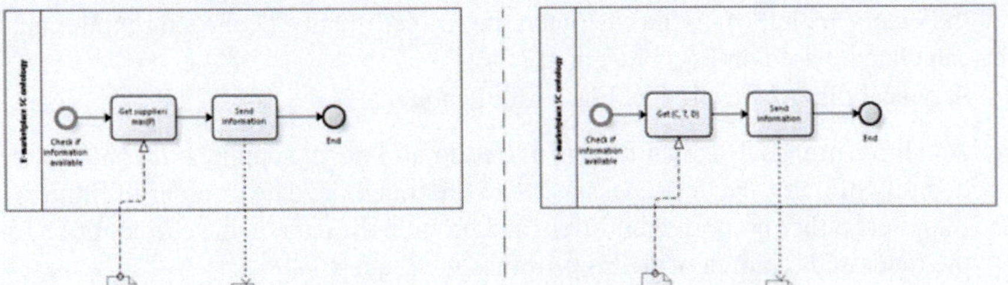

Fig. 17.5 Supply chain node's agent registration [12]

When an agent of a type *Supplier* is being registered, it is enlisted with its respective products or services in the *yellow-page service*. Supplier-agent's related *QoS indicators* are saved into its SupplierContentMessage object:

- Correctness: average value of the Correctness indicator for all business transactions made in its history of participation in the supply chain;
- Timeliness: average value of the Timeliness indicator for all business transactions made in its history of participation in the supply chain;
- Dependability: average value of the Dependability indicator for all business transactions made in its history of participation in the supply chain;
- Price: value of the currently offered price for the specific item;
- Name: unique identifier of the supplier.

When an agent of a type *Customer* is being registered in the *yellow-page service* it is enlisted by its name. Upon its first order, the query to the e-marketplace ontology is made for all eligible suppliers in the supply chain. Based on that query, additional customer-related *QoS indicators* are saved into the CustomerKnowledge object:

- Name: unique identifier of the customer;
- Price: maximum price from all customer suppliers' business transactions.

After every transaction these values are updated correspondingly.

When a specific proposal is made by a supplier, a serialized object SupplierContentMessage, is contained in the suppliers' message. Based on that knowledge from suppliers (average values of the C, T, D indicators) and knowledge about a specific supplier, stored in the CustomerKnowledge object (offered price P and maximum price $max\,P$), the overall price indicator is calculated based on Eq. (17.6). As by now all indicators are known to the customer, the service quality Q of each supplier may be calculated, based on Eq. (17.7) or (17.8).

A typical ordering cycle (cp. Fig. 17.6) then reads:

1. A call for proposals for an item is posted to all known suppliers; it contains the quantity of items the customer wishes to order and the date of order fulfillment.
2. Suppliers either propose their offers and provide the data on the current price for the requested (amount of) items or refuse the request.
3. The customer makes the decision on the acceptance of a certain proposal based on the prices (P) and the service quality indicators (Q) of all suppliers who provided positive feedback to the inquiry.
4. An order is then made to the selected supplier by posting it an accept-proposal, while a reject-proposal is sent to all other participating suppliers.
5. The order is either acknowledged as being fulfilled or a failure to fulfill the order is issued by the chosen supplier.
6. Based on the outcome of the inquiry, service quality indicators of the supplier-nodes are recalculated to be considered with the next ordering cycle.

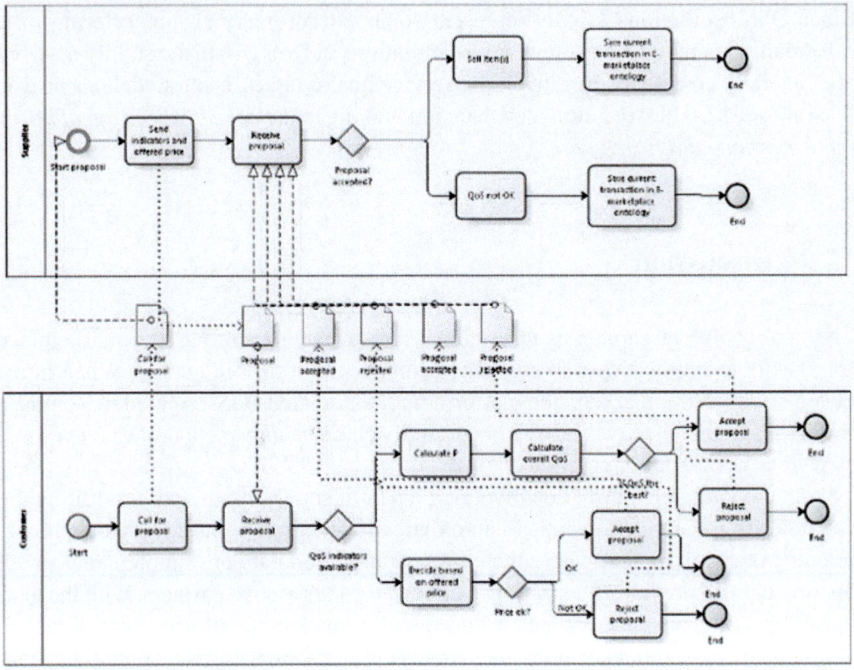

Fig. 17.6 Ordering cycle [12]

17.4 Evaluation

In our experiment [12, 15], we have analyzed the effect of *QoS oriented decision making* vs. minimum price and empirical decision making on supply chain management. We could establish that decisions made solely based on the price (P) criterion could be misleading. On the other hand, decisions made only based on experience could be sub-optimal. In many cases Q as a decision criterion would match their combined effect. Hence, the introduction of the Bayesian model, where every new Q represents a gain QG (positive or negative) to the current value of Q, renders an impartial and sustainable merit of QoS within the supply network.

From the analysis, we could conclude that the Bayesian service quality evaluation model generally has a positive effect on overall supply chain service quality of the observed supply chain. Considering the values of the quality indicators Q_p and Q, a claim can be made that the decision model based on service quality would positively influence supply chain management business logic for two reasons:

1. it prevents making quick uninformed decisions based on price variations.
2. considering the supply chain as a whole, the ordering decisions based on QoS result in its gradual improvement and lower oscillations.

Since QoS oscillations usually represent some extraordinary events, offering little information on their background, high oscillations in QoS within the supply network are generally considered negative. The service quality assessment model has proven feasible and suitable for automated decision making in the *decentralized agent based e-marketplace platform* [12].

17.5 Discussion

The goal of this research was to establish a distributed agent based e-marketplace framework to improve overall company's supply chain service quality. When introduced as a global e-marketplace platform with automated QoS-oriented reasoning it would render an objective and universal supply chain management broker for vertical as well as horizontal integrations.

Any e-commerce transaction involving a pair of supply chain nodes, fulfilling the customer-supplier relationship, is automatically integrated into a service-quality-oriented decision model, where the current transaction data are supplemented with historic data on previous transactions between the perspective partners, with the goal of best meeting customer's needs. Consequently, since more informed decisions are taking place autonomously along the whole supply chain, its service quality is being improved up- and downstream. In contrast to centralized e-marketplace platforms, the agent-based approach is distributed and dynamic, and hence corresponds with supply chains' natural behavior.

To apply the decentralized agent-based approach to supply chain management, any pair of supply chain partners are autonomously integrated in both—cyber and physical—spaces with the goal of meeting supply chain nodes' service quality goals. Considering the service quality criteria, by representing a supply chain as a multi-agent system and making supply chain operation decisions according to this service quality assessment model, up to 50% better informed decisions can be achieved, resulting in higher customer satisfaction and more sustainable supply chains.

17.6 Conclusion

In supply chain management two approaches to service transformation in the digital era can be observed for efficient, service quality-oriented supply chain management:

1. Centralized: e-marketplace web portals, combining demand and supply in a temporally and spatially shared repository and
2. Decentralized: agent-based e-marketplaces with a yellow-page dictionary service and distributed network of temporally and spatially distributed agents of supply chain partners.

In this research the latter approach has been chosen, since it resembles the natural behavior of supply chain nodes, which: appear, register their services, handle inquiries and orders and establish their own service quality-based supply chains on the supply chain network with their partners. Here, the structure, behavior, and communication models of the proposed multi-agent framework with a service quality oriented assessment model for collaboration within supply chains have been presented. An underlying e-commerce ontology has been introduced. It introduced the different categories of nodes and defined the valid communication message contents. The method of assessing the service quality and its adoption into supply chain management operations has been introduced into the knowledge base, maintained by the ontology in the form of a Yellow page service. Finally, the service-quality-based supply chain management model has been statistically evaluated against randomly created e-commerce transactions. The most important finding of this research is that the service quality decision model adds more than 50% of information when choosing a supply chain partner for collaboration on the customer side (i.e. as opposed to decisions based solely on product/service price). In addition to the knowledge on the supply network stored in the ontology, by *learning processes* additional business logic can be embedded in agents (cp. Chap. 18).

By adopting the decentralized agent-based e-marketplace platform the service quality within supply chains can be significantly improved. Since supply chain partners don't need to disclose any additional information on their business transactions than they already do with their current supply chain management operations, there is also no reason for them to reject the proposed solution due to information hiding.

References

1. Scholz-Reiter B, Sowade S, Rippel D (2011) Modeling the infrastructure of autonomous logistic control systems. In: Proceedings of the European conference of systems, and European conference of circuits technology and devices, and European conference of communications, and European conference on Computer science, vol 5
2. Hulschman M, Windt K (eds) (2007) Understanding of autonomous cooperation and control in logistics: the impact of autonomy on management, information. communication and material flow. Springer, Berlin
3. Saaty T (2008) Relative measurement and its generalization in decision making why pairwise comparisons are central in mathematics for the measurement of intangible factors the analytic hierarchy/network process. Revista de la Real Academia de Ciencias Exactas, Fisicas y Naturales - Serie A: Matematicas 102:251–318
4. Fox M, Barbuceanu M, Teigen R (2000) Agent-oriented supply-chain management. Int J Flex Manuf Syst 12(2–3):165–188
5. Sowa J (2000) Knowledge representation: logical, philosophical, and computational foundations. Brooks Publishing Co., Pacific Grove
6. Scholz-Reiter B, Hohns H, Kolditz J, Hildebrandt T (2005) Autonomous supply net coordination. In: In proceedings of the 38th CIRP manufacturing systems seminar, Florianopolis, Brazil
7. Oestereich B, Weiss C, Schroder C (eds) (2003) Objektorientierte Geschaftsprozessmodellierung mit der UML. dpunkt verlag, Heidelberg

8. Weiss G, Jakob R (2005) Agentenorientierte Softwareentwicklung
9. Gumzej R, Sukjit P, Unger H (2012) Modeling overlay networks for autonomous supply chain systems. Log & Sust Trans Celje, Slovenia 3:41–50
10. Russell S, Norvig P (2009) Artificial Intelligence: A Modern Approach, 3rd edn. Prentice Hall Press, Hoboken
11. Fensel D (2001) Ontologies, pp 11–18. Springer, Berlin. https://doi.org/10.1007/978-3-662-04396-7_2
12. Milić B, Rosi B, Gumzej R (2019) An approach to e-marketplace automation 26(3):639–649. https://doi.org/10.17559/TV-20171201150248
13. Bellifemine FL, Caire G, Greenwood D (2007) Developing multi-agent systems with JADE (Wiley Series in Agent Technology). Wiley, Hoboken
14. Gumzej R, Gajšek B (2011) A virtual supply chain model for qos assessment, vol 3. Springer, Berlin, pp 147–157
15. Poletan Jugović T, Čišić D, Gumzej R (2019) Supply chain service quality improvement by e-marketplace automation 31(2):185–194. https://doi.org/10.7307/ptt.v31i2.3042. https://traffic.fpz.hr/index.php/PROMTT/article/view/3042
16. Bellifemine F, Poggi A, Rimassa G (2001) Developing multi-agent systems with a fipa-compliant agent framework. Softw. Pract. Exper. 31:103–128. https://doi.org/10.1002/1097-024X(200102)31:2<103::AID-SPE358>3.0.CO;2-O
17. FIPA: Foundation for intelligent, physical agents. http://www.fipa.org/
18. Foundation for Intelligent Physical Agents: Fipa 97 specification - part 2 - agent communication language. http://www.fipa.org/specs/fipa00018/OC00018.pdf

Chapter 18
Use Case: E-Marketplace Regulation

Problem 18.1 As networks of organizations, supply chains operate in a partly regulated environment with clearly defined objectives. In Chap. 17 a digital framework of a supply network environment has been presented, together with the rules of engagement and supply network knowledge management and exploration as a means to achieving the goals of individual supply chain echelons. Although each individual node shall pursue its own goals, such a loosely coupled system requires some form of regulation to fulfil the goals of the system. How to model supply chains and how to achieve the goals of a *supply chain regulation*?

Solution 18.1 In a partly regulated environment, defining only the legal transactions among supply chain echelons the goals are left to them to achieve by whatever means they deem appropriate. On one hand the goal of echelons may be to maximize their profits, while on the other and in the long run it is in their best interest to maximize their service quality as well. What is the role of market regulators in achieving and maintaining the trustworthiness of the participants and how to include it in the digital framework? Their role is to maintain the principles of "fairness" in a partly regulated marketplace. In order to do so, they may introduce rewards, fines and other regulatory measures they deem appropriate. Are they really necessary or will the market regulate itself? As their primary goal differs from the goals of other echelons in the supply network, they represent another kind of agents. Can they rely on existing knowledge? How to model their behaviour and integrate them in the digital framework of our supply network? How does their operation affect other participants in the network? This use case will provide answers to these questions.

18.1 Introduction

The *supply chain* is a network of organizations working together in different processes and activities in order to bring products and services to the market, with the purpose of satisfying customer's demands [1]. The fulfillment of its objectives is accomplished

R. Gumzej, *Intelligent Logistics Systems for Smart Cities and Communities*, Lecture Notes in Intelligent Transportation and Infrastructure,
https://doi.org/10.1007/978-3-030-81203-4_18

by three types of contributing parties: (1) *suppliers*, (2) *customers* and (3) *market regulators*. The main role of the latter is to set the rules of engagement and monitor their fulfillment to prevent fraud as well as minimize the probability of dissatisfaction of customers when dealing with distrustful suppliers.

A supply chain can be seen as a complex adaptive system of interrelated and interdependent elements, which collaborate in order to achieve some preliminary set objectives. A *complex adaptive system* is a system, which evolves through time into a coherent whole that adjusts and *self-organizes* itself without any individual entity intentionally controlling or supervising it [2].

Decision makers in supply chains are often faced with strategic decision-making challenges with only partial information on system structure and/or operation. Hence, the adopted decisions may have a negative impact on overall supply chain performance. These situations should be avoided and can be prevented by utilizing *Supply Chain Management* (*SCM*) platforms and *Decision Support Tools* (*DST*), as indicated in Chap. 17. SCM addresses a wide range of problems, which may be divided into three categories [3]:

1. supply chain infrastructure (network) design,
2. supply chain analysis and policy formulation, and
3. supply chain planning and scheduling.

As indicated above, the main objective of SCM is the maximization of total profits derived across different echelons of the supply chain, while its second and most important objective is to maximize customer satisfaction, expressed by the service level [4], also referred to as Quality of Service (QoS) in the related chapters. While the majority of existing decision support tools support the first objective, the fulfillment of the second objective is interrelated with it and should as such be considered in conjunction [5]. Its goals can only be attained by adequate supply chain configuration and proper use of management tools. The latter are provided by SCM, representing the process of planning, implementing and controlling the operations of the supply chain in an efficient way. SCM spans all movements and storage of raw materials, work-in-process inventory, and finished goods from the point of origin to the point of consumption [6].

DSTs enable business and/or organizational decision-making based on rapidly changing data not known in advance. Within the process of modeling and solving the indicated problems, managers have to make decisions on different hierarchical levels. These decisions can be classified as strategic, tactical and operational, depending on their effects to the overall supply chain. Due to interdependence among the three levels, SCM remains incapable of satisfactorily addressing many practical real-world problems at a strategic level. Let alone that these levels are difficult to tackle at any individual level, due to inter-dependencies among entities and their autonomous behavior they are even more difficult to address. Hence, the majority of modeling approaches in DST provide invalid results [7]. For obtaining adequate solutions at the strategic level, decision-makers need comprehensive models to guide them in the decision-making process. In our digitized supply chain the behavior of the modeled system depends on decisions and interactions among heterogeneous entities, represented by agents. Hence, the *Agent-Based Simulation* (*ABS*) approach to modeling and simulation of such systems may be used to construct a consistent and coherent DST model and allow for knowledge based reasoning, as established by the decentralized agent based e-marketplace platform, introduced in Chap. 17.

Applying the holistic modeling approach to capture the extended supply chain, involving all three types of the previously mentioned parties, within a DST at a strategic level may provide a solution to our supply chain regulation problem. Here, an ABS-based model of a multi-echelon supply chain, which may be used to analyze and fine tune the behavior of supply chain nodes, is presented. First, the digitized supply chain model and its integration into an ABS model are presented. Next, the ABS model of a generic supply chain is presented with the demonstration of the nodes behavior according to the selected strategy. Finally, the results of different simulation scenarios are analyzed and a conclusion on the outcomes is drawn.

18.2 Service Quality Model

Our supply chain evaluation method is based on the adaptable (customer-led) supply chain model and considers the supply chain as a composition of partners, fulfilling the customer-supplier relationship. QoS-probes are inserted on every link among any two elements fulfilling this relation. Hereby, service quality is observed between producers and their suppliers of materials and components, producers and retailers of their products, as well as customers and retailers. Any input transaction on a link represents a response to a call for proposals and has two properties—price and QoS value—that are considered by a customer when deciding on the acceptance or refusal of a supplier's proposal. Currently, the decisions are made solely based on price or preference, where the lowest price or highest preference is considered the best offer. However, due to disregard of QoS such decisions are not always the best ones. Customers usually rely on a (set of) trusted supplier(s) with whom they establish contract relationships that are considered a priority above the price criterion. This criterion is QoS-related and represents a level of trust among a customer and its supplier. Due to maximization of suppliers profits they may sometimes choose to employ practices that would lower their QoS and would normally represent a void of the established trust.

Prior to any transaction between supply chain partners the corresponding probe is evaluated based on the provided values for the individual partner's QoS and price. With each transaction the partner's QoS value is updated and stored for later reference. The gathered data may be used for monitoring and analysis of supply chain excellence, represented by the mean value of all partner's QoS values. In order for this benchmark to be meaningful, all measured values must lie within a predefined threshold. Hence, all QoS indicators are normalized and joined in a single value, representing QoS of any distinct node in the supply chain. It shall be used to enable quick decisions and comparison among supply chains by their partners. By constructing and embedding an QoS-based agent based simulation model in a DST system, closed loop analysis and performing "what-if" scenarios are enabled to assist the decision makers at supply chain nodes' with their informed SCM decisions.

18.3 Supply Chain Model

Since business relations are essentially about obtaining and maintaining supply chain partner's wealth, in this research the effect of good and bad business practices is reflected by positive and negative financial stimuli. Market regulators in the form of inspectors are introduced into the model to monitor the state of trustfulness (QoS) of supply chain partners. An agent-based simulation model has been devised to monitor the overall QoS within a supply chain, based on the given initial number of producers, retailers and market regulators as well as the parameters that represent the market regulation policy. By different scenarios the most efficient distribution of contractual and non-contractual partners as well as market regulation policy can be discovered. Since the geographical distribution of the producers and retailers in space is unknown, we assume it to be random. Hence, also their positioning does not affect their pricing policies in any way.

The framework of the model consists of three agent types:

1. producer agents,
2. retailer agents, and
3. regulator agents

where there are n producers in a product production base, and each producer produces s units of a product. At a wholesale market, there are m retailers that deal with the producers. There are x regulator agents who perform inspections on producers and retailers. The behaviors of the different types of agents are further described below.

Considering the trust-related credence attributes between customers and suppliers, there is a problem of information asymmetry, since they cannot be directly observed and are difficult to measure. The producers can be trusted or distrusted and it depends on their decision regarding their quality of service (QoS). The QoS value is calculated based on the given model (cp. Chap. 17) to determine, whether a customer should establish or disrupt contract relationships with its suppliers. Market regulators intervene with inspections throughout supply chains in order to assure an adequate level of quality of products and/or services offered. Products or services of lower quality are deprecated, since they are resulting in lowering partner's wealth and usually also result in the loss of trust. Hence, the tendency is to build contract relations only among trustworthy supply chain partners. Only by this the overall QoS of a supply chain may be improved and result in a sustainable SCM strategy. In our supply chain model, non-contractual retailers make transactions with non-contractual producers sporadically. Contractual retailers make transactions with contractual producers building fixed transaction relationships. Contractual modes are trust-related and can be switched under certain circumstances. Producers' and retailers' trustfulness is monitored by the regulator agents, who issue fines in case inadequate business practices are detected.

18.3.1 Regulator Agent

In the model, the regulator agents are several market regulator's departments which supervise different parts of the supply chain. Here, for simplicity reasons, we assume two types of regulator agents, one for producers and another for retailers, respectively. The goal of the market regulators is to improve the service quality within the supply chain by means of sampling inspection and punishment under certain conditions. The goal of the retailers is to maintain contract relationships with trusted producers only. There are different inspecting ratios for producers Θ_1 and retailers Θ_2. α is the frequency of product inspections. The fine on producers is f_1, and the fine on retailers is f_2. They are understood per product shipped or product/service sold.

18.3.2 Producer Agent

A producer produces a number of products and sells them to one of the retailers. The producer's goal is to maximize its profit. The attributes of a producer are producer's type (contractual or non-contractual), trustfulness, wealth, and transformation threshold.

The producer's *trustfulness level* (QoS) indicates its ability to resist the shock of benefit produced by the gap between expected returns of trusted and distrusted production. The decision on trustfulness is built into the model as a function of producers' experience. With every transaction the producer gains experience with regard to its benefit or loss, in case during inspection its inadequate business practice was detected. The following formulas show quality of service updating as a function of producer's experience under different situations:

$$QoS_{t+1} = (1\text{–}d^{+}) * QoS_t + d^{+} * QoS_{\min} \tag{18.1}$$

$$QoS_{t+1} = (1\text{–}d^{-}) * QoS_t + QoS_{\min} \tag{18.2}$$

where QoS_t represents the value of QoS after t transactions. A value of QoS close to 0 represents the fact that the producer is not trustworthy, while a QoS value close to 1 represents its commitment to be trustworthy. The parameters d^{+} and d^{-} where $(0 < d^{+} < 1, 0 < d^{-} < 1, d^{-} > d^{+})$ represent impact factors of positive and negative experience, respectively. The parameter QoS_{min} represents a minimal level of trust, where $(0 < QoS_{min} < 1)$ and is initially set to 0.1. If the producer is not found cheating, its new QoS value is calculated according to (18.1), otherwise the new value is calculated according to (18.2).

The updating of the producer's experience and hence the decision on the investment into a higher QoS production is simplified. It is based on the assumption that the most recent experience has the stronger influence, representing the short memory effect. It is also influenced by the fact that a negative experience has a stronger influence than a positive experience, representing the endowment effect.

The producer sells its products to the retailers after production, and makes a profit that equals to total revenue minus total cost. The revenue is the sales revenue from the market, and the cost includes the production cost and a fine, if producer's ill-practices have been detected. Furthermore, the profit changes the producer's wealth. We assume a uniform transaction price ($p1$) produced when the transaction between a producer and a retailer happens:

$$p_1(t) = \lambda cs + (1 - \lambda)cl + \varepsilon(t) \tag{18.3}$$

where cs is the cost of trusted production for a non-contract producer, cl is the cost of distrusted production. $\varepsilon(t)$ is a random real number following normal distribution $N(\mu, \sigma 2)$, which reflects fluctuations in the market. λ is a weight set in the interval $(0, 1)$, representing market flexibility.

The producer's decision on trustworthy production mainly depends on three parameters: economic benefits comparison, random factors and current level of QoS. The following formula expresses this production decision:

$$\lambda k_i + (1 - \lambda)\rho > QoS_t \tag{18.4}$$

If this inequality is satisfied, the producer distributes a certain number of products in an inadequate way, hereby reducing its QoS by QoS_{min}. Otherwise, the producer distributes a certain number of products in an adequate way, hereby increasing its QoS by QoS_{min}. Here ρ is a random number $(0 < \rho < 1)$; k_i is a normalized value, which is given based on the gap between expected returns by trusted production, which may be calculated in two ways based on the fact whether the producer is a non-contractual producer (18.5) or contractual producer (18.6):

$$k_1 = (cs - cl - f_1/n_1)/(cs - cl) \tag{18.5}$$

$$k_2 = [c's{-}cl - f_1\Theta_1/n_1 - f_3\alpha\Theta_2/m_1bs + (f_1 + f_3)\alpha\Theta_1\Theta_2/n_1m_1bs]/(c's{-}cl) \tag{18.6}$$

where $c's$ is the cost of trusted production for a contractual producer, $f3$ is a fine deposit made by contract producers to the retailers for any product sold.

Each producer can decide whether to make a contract with a retailer. If a producer makes a contract with a retailer, the producer will become a contract producer, and vice versa. The non-contract producers sell their products to the nearest non-contract retailers. Each producer with non-trustworthy production is fined, if the producer is inspected. The contract producers sell their products to the retailers who have contracts with them. The contract producer's fine mainly derives from two aspects: inspections and investigating and tracing of its contractual retailers. It is assumed that the contract retailers can find and fine their distrusted contract producers easily. For s products sold the producers' profits may be calculated as follows.

1. Non-contract producers:

   a. $pi_1 = s(p_1 - cl)$, when the producer is distrusted and is not inspected;
   b. $pi_2 = s(p_1 - cl - f_1)$, when the producer is distrusted and fails to pass the inspection;
   c. $pi_3 = s(p_1 - cs)$, when the producer is trusted;

2. Contract producers:

   - When the producer is distrusted and transaction price between producer and retailer is more than minimum protective price ($p1 \geq pr$), the profit is:

     a. $pi_1 = s(p_1 - cl)$, if the producer is not fined;
     b. $pi_2 = s(p_1 - cl - f_1)$, if the producer does not pass the inspection conducted by the market regulator;
     c. $pi_3 = s(p_1 - cl - f_3)$, if the producer is fined by one's own contract retailer;
     d. $pi_4 = s(p_1 - cl - f_1 - f3)$, if the producer is fined by the market regulator and its contract retailer.

   - When the producer is trusted and ($p_1 \geq pr$), the profit is $pi_5 = s(p_1 - c's)$.
   - When the producer is distrusted and ($p_1 < pr$), the profit is:

     a. $pi_6 = s(pr - cl)$, if the producer is not fined;
     b. $pi_7 = s(pr - cl - f_1)$, if the producer does not pass the inspection conducted by the market regulator;
     c. $pi_8 = s(pr - cl - f_3)$, if the producer is fined by its contract retailer;
     d. $pi_9 = s(pr - cl - f_1 - f_3)$, if the producer is fined by the market regulator and the contract retailer.

   - When the producer's production is trusted and ($p_1 < pr$), the profit is: $pi_{10} = s(pr - c's)$.

Considering the learning capacity in this model, producers can learn and imitate from each other. If the non-contract producer's profit is less than the average profit of its neighbors, who are contract producers, it would transform into a contract producer under certain conditions. Otherwise, its type remains the same. The contract producer is in a similar situation. If the contract producer's profit is less than the average profit of its neighbors, who are non-contract producers, the contract producer would transform into a non-contract producer under certain conditions. Otherwise, the contract producer would sustain its contracts with its retailers.

Different types of producers usually coexist simultaneously in a simulation run. Prior to transformation any producer needs to take into account the financial consequences of its transformation. When the difference between the benefits and costs of the transformation exceeds a threshold γ (18.7) the producer would change from one type to another, while otherwise its type would remain the same:

$$\lambda(pi'i - pii)/pi'i + (1 - \lambda)\rho > \gamma \tag{18.7}$$

where pii is producer(i)'s profit, $pi'i$ is the average profit of its neighbors, who are a different type of producers, parameter λ is a weight set in the interval (0, 1). Parameter ρ is a random number ($0 < \rho < 1$) and reflects fluctuations in the market.

Each producer has its own transforming threshold that implies producers' adaptability to the environment, production decision regarding its QoS and experience gained by it. The producer's transforming threshold γ updates with every transaction according to (18.8), if the producer is not fined, and according to (18.9) if the producer is fined:

$$\gamma_{t+1} = \gamma_t(1 - QoS_t) + \gamma_t \tag{18.8}$$

$$\gamma_{t+1} = \gamma_t(1 - QoS_t) \tag{18.9}$$

18.3.3 Retailer Agent

A retailer purchases products from producers in different ways, and sells them on to those downstream customers in the supply chain. The retailer's goal is to gain more profit. The attributes of a retailer are contract status (non-contracted or contracted), wealth value, and transforming threshold. The behavior of a retailer follows the rules described below. The retailer sells the products to the downstream of supply chain after purchase, and makes a profit that equals to the total revenue minus the total cost, where revenue represents sales revenue from the market, and the cost the purchasing cost plus the fine deposit. Furthermore, the profit makes the retailer's wealth changing.

We assume that each retailer sells all products, regardless of the type of producers, and that there is a uniform transaction cost (p_2) produced when the transition between the retailer and the downstream customer is done.

$$p_2(t) = p_1(t)(1 + \delta) \tag{18.10}$$

where parameter δ represents the extra ratio between wholesale price and market price and is a real number from the interval (0, 1). p_2 fluctuates according to p_1 on a certain scale.

When contract retailers make a contract with contract producers they provide a minimum protective price (pr). When market price p_1 is higher than the protective price, the purchasing price fluctuates along with the market changes. When market price p_1 is lower than protective price, the purchasing price becomes the protective price. Retailer's profits may be calculated as given below, where bs denotes the numbers of products sold.

1. Non-contract retailer:

   a. When the retailer is not inspected by the market regulator, the profit is: $ri_1 = bs(p_2 - p_1)$.
   b. When the retailer does not pass the inspection, the profit is: $ri_2 = bs(p_2 - p_1 - f_2)$.

2. Contract retailer:

   - When $p_1 \geq pr$, the profit is:

     a. $ri_1 = bs(p_2 - p_1)$, if the retailer is not fined;
     b. $ri_2 = bs(p_2 - p_1 - f_2) - ci + f_3$, if the retailer is fined by the market regulator;

   - When p1 < pr, the profit is:

     a. $ri_3 = bs(p_2 - pr)$, if the retailer is not fined;
     b. $ri_4 = bs(p_2 - pr - f2) - ci + f_3$, if the retailer is fined by the market regulator.

where q is the number of contract producers who are inspected by their contract retailers as a consequence of them being fined by the market regulator.

For the non-contract retailers, there is no way to find out about distrusted producers, if they fail the inspection, because they sporadically interact with some of a large number of small-scale producers in the market. Consequently, the non-contract retailers alone have to bear the fine from the market regulator.

On the other hand, the contract retailers can find those distrusted producers, if they fail inspection, by several means (such as traceability system, field management and/or certification). When a contract retailer is fined, it would conduct an investigation about the distrusted producers at a cost (ci), and punish those contract producers. Their punishment is a fine in the form of a deposit (f_3) that the contract producers have paid in advance. Contract retailers build up stable interaction relationships with their producers. Moreover, their trust is strengthened with every solid transaction, resulting in their rising QoS.

In this model, the two types of retailers can learn and imitate from each other. When the non-contract retailer's profit is less than the average profit of the contract retailers, it would transform into a contract retailer under certain conditions. Otherwise, its contract status would remain the same. The contract retailer is in a similar situation and would transform into a non-contract retailer under certain conditions.

Different types of retailers usually coexist simultaneously in a simulation run. Prior to transformation any retailer needs to take into account the financial consequences of its transformation. When the difference between the benefits and costs of the transformation exceeds a threshold ν (18.11) the retailer would change from one type to another, while otherwise its type would remain the same.

$$\lambda(ri'i - rii)/ri'i + (1 - \lambda)\rho > \nu \tag{18.11}$$

where rii is retailer(i)'s profit, and $ri'i$ is the average profit of the retailers who are of different type, parameter λ is a weight set in the interval $(0, 1)$. Parameter ρ is a random number from the interval $(0 < \rho < 1)$ and reflects fluctuations in the market.

Each retailer has its own transforming threshold that implies retailers' adaptability to the environment, sales decision regarding its QoS and experience gained by it. The retailer's transforming threshold ν updates with every transaction according to (18.12) if the seller is not fined and according to (18.13) if the seller is fined:

$$\nu_{t+1} = \nu_t(1 - QoS_t) + \nu_t \tag{18.12}$$

$$\nu_{t+1} = \nu_t(1 - QoS_t) \tag{18.13}$$

18.4 Simulation

Our DST was developed and tested with NetLogo 5.3.1 [8]. The simulation model applies to generic supply chains of partners fulfilling the supplier-customer relationship. In our experiment [9] we have focused on producers of final products and retailers of these products, who sell them to end-customers. Our goal was to study the introduction of QoS into the SCM strategy. In our case QoS represents supply chain partner's trustfulness as a gradual quantitative indicator that changes (improves or deteriorates) with each transaction. The results of this oscillations are considered as benefits or hazards when making business deals and while considering contract relationship status transitions between business partners. With our model, we have been able to evaluate the combined impact of a number of contractual and/or non-contractual producers and retailers on the market, who are inspected by a number of market regulators with the given inspection frequencies and different fines on non-trustworthy business practices of producers and/or retailers. Hereby, we have also observed the fluctuations in the market, regarding the number of contractual relations, dynamics of QoS changes and qualified rate of trustful partners' changes.

The input variables of our simulation-based DST (cp. Fig. 18.1) are:

- n number of producers,
- m number of retailers,
- x number of inspectors,
- f_1 producer's fine per product,
- f_2 retailer's fine per product,
- f_3 fine deposit by contracted producers,
- Θ_1 inspection ratio for producers,

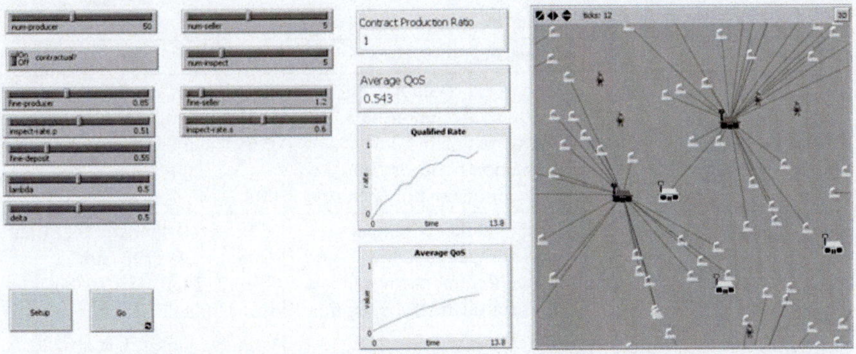

Fig. 18.1 Agent based simulation model

Θ_2 inspection ratio for retailers,
λ flexibility of the market,
δ ratio between wholesale price and market price.

18.5 Evaluation

Table 18.1 summarizes the typical scenarios that were analyzed in the simulation. They pertain to different kinds of markets ranging from a free markets with no contractual relationships and low regulation over to highly regulated markets where contractual relationships dominate the market. In all scenarios the achieved QoS has been observed. From the obtained results in general a claim can be made that QoS is a good market regulator, since customers not only tend to build relations with trustful suppliers but are also interested in themselves maintaining a high QoS level. While the model may be used as a DST in SCM operations, it also provides for a clue on how to build the SCM business logic into the operations of the agents representing supply chain echelons in our digitized supp network. A more elaborate presentation of the outcomes is given in the discussion below.

18.6 Discussion

Question 18.1 How does the introduction of QoS impact our marketplace?

Cliam 18.1 The results of the simulation [9] rendered the fact that QoS impact on the marketplace is substantial. The producers as well as retailers are striving to strengthen their position in the marketplace. With every transaction their QoS is updated, which may lead to a strategic decision to change their current contract

Table 18.1 Results obtained from four main scenarios

Scenario	Properties	Parameters
1	Free market with no contractual relationships, few market regulators, low inspection frequency and low fines for distrustful operation	C/P: avg. 0; std. 0 QoS: avg. 0,578; std. 0,02044 QR: avg. 0,695; std. 0,03837
2	Mixed market with many market regulators, low inspection frequency and low fines for distrustful operation	C/P: avg. 0,494; std. 0,521029 QoS: avg. 0,5798; std. 0,021311 QR: avg. 0,7037; std. 0,060802
3	Mixed market with many market regulators, high inspection frequency and high fines for distrustful operation	C/P: avg. 0,9; std. 0,316228 QoS: avg. 0,725; std. 0,019579 QR: avg. 0,887; std. 0,023594
4	Mixed market with few market regulators, high inspection frequency and low fines for distrustful operation	C/P: avg. 0,7; std. 0,483046 QoS: avg. 0,728; std. 0,016193 QR: avg. 0,85; std. 0,026247
C/P (contract production)	QoS (Quality of Service)	QR (Qualified Rate)

status in case they discover that their profits would increase with the status change. Their decision depends on the threshold that is affected by their experience and QoS values. With every transaction resulting in their positive/negative experience their QoS is increased or decreased accordingly and also affects their future behavior. In case their profit is above/below average profits of their neighbors and their threshold is low, their contract status may change.

Question 18.2 How do fines and the ratio (frequency) of inspections affect our marketplace?

Cliam 18.2 Fines from the market regulators largely affect the marketplace. However, to be effective, the frequency of inspections needs to be high, which is associated with a corresponding number of inspectors. According to the results, we may conclude that the frequencies of inspections have a bigger impact on the market than the amounts of fines.

Question 18.3 By which combination of measures can we obtain a higher qualified rate of business partners' QoS?

Cliam 18.3 The results have shown that in all scenarios there is an initial "warm-up" period of some 30 transactions when QoS grows from its minimum to its average value. After that it does not change much. Hence, a conclusion can be made that QoS stabilizes the market. The qualified rate (QR) represents a share of business partners whose QoS is above the average QoS. During the growth of QoS the QR's oscillations are high, since the model has not yet reached a stable state. Once the

QoS reaches its average value, however, the QR indicator also reaches a certain level and its oscillations are lower. The remaining oscillations are mostly associated with partners changing their contract status and hence increasing or decreasing their exposure in the market.

Question 18.4 By which combination of measures can we reach a situation, when the market is totally regulated (all partners are trustworthy contract partners)?

Cliam 18.4 On a fully regulated market one would expect the QoS and QR rates to nearly reach 1, since all partners are considered trustful. However, since they still compete in the market this value is never reached. In the rest of the cases the expected behavior could be observed in a free or mixed market.

Considering all the above claims, we may conclude that the market is largely affected by a combination of the following three parameters: number of market regulators, fines and inspection rates. In case they are high, we are likely to end up with a fully regulated market. Otherwise, it is more likely that we end up with a mixed or even a free market.

18.7 Conclusion

QoS is considered a good market regulation criterion – even in free markets the competition among non-contract partners renders an above-average overall QoS. Of course one may reach a higher QoS with contract partners in case their goal is trustworthiness. To maintain an elevated level of QoS, inspections are necessary to prevent bad business practices to prevail due to higher expected profits. On the other hand, market inspections only have a significant impact on the model, in case they are frequent. High fines for distrustful business practices alone are not sufficient market regulators.

In this chapter not only the business logic behind the strategic decisions of supply chain echelons has been integrated into the supply chain model to serve as a DST, the role of the market regulators and their affect on the market have been modeled as well. *Supplier* agents have been modeled as producers. *Customer* agents have been modeled as sellers. *Market regulator* agents have been modeled as inspectors. Since their goals are different but related in the proposed decentralized agent based e-marketplace framework (cp. Chap. 17), the results shall serve as basis on how to enhance the knowledge base of the model-based learning agents used. Apart from supply chain procedures and knowledge already contained in the *yellow-page service*, the knowledge obtained from this example shall serve as business logic model of the behavior of supplier and customer agents.

As a conclusion from this and Chap. 17 we may claim that the *QoS* criteria support supply chain partners with their informed decisions in SCM operations and hereby regulate their supply chains.

References

1. Christoper M (2005) Logistics and supply chain management. Prentice Hall, London
2. Holland JH (1995) Hidden order: how adaptation builds complexity. Helix books: science. Basic books. https://books.google.si/books?id=jQHvAAAAMAAJ
3. Shah N (2005) Process industry supply chains: advances and challenges. Comput Chem Eng 29(6):1225–1236
4. Zhang G, Shang J, Li W (2011) Collaborative production planning of supply chain under price and demand uncertainty. Eur J Oper Res 215(3):590–603
5. Gumzej R, Gajšek B (2011) A virtual supply chain model for qos assesment. Springer Verlag Berlin Heidelberg, vol. 3, pp 147–157
6. Melo M, Nickel S, Saldanha-da Gama F (2009) Facility location and supply chain management—a review. Eur J Oper Res 196(2):401–412
7. Georgiadis P, Vlachos D, Iakovou E (2005) A system dynamics modeling framework for the strategic supply chain management of food chains. J Food Eng 70(3):351–364
8. Wilensky U Netlogo. http://ccl.northwestern.edu/netlogo/
9. Gumzej R, Rosi B (2017a) An agent-based simulation of a qos-oriented supply chain **29**(6), 593–601. https://doi.org/10.7307/ptt.v29i6.2520, https://traffic.fpz.hr/index.php/PROMTT/article/view/2520

Chapter 19
Use Case: Intelligent Transport Unit

Check for updates

Problem 19.1 How to design an *Intelligent Transport Unit* (*iTU*) capable of autonomous routing throughout the *Physical Internet* (*PhI*)?

Solution 19.1 From a physical perspective, an iTU (cp. Fig. 19.1) must be easy to handle, store, transport, seal, snap to a structure, interlock, load, unload, build and recycle (green logistics)—just like all *PhI Containers* [1]. *Logistic hubs* (e.g. warehouses) utilize fixed and mobile RFID transceivers to monitor passing cargo units from the point they arrive, during their storage, and manipulation, until they leave the hub. A GPS receiver, installed at the monitoring station, can delegate exact position information to the stored or conveyed iTUs. This way, mobile iTUs can maintain accurate position data, even in closed spaces such as warehouses, or during transit, even without themselves being equipped with a GPS receiver.

19.1 Introduction

In Logistics 4.0 greater transparency of the logistic processes is required, as the customers not only demand to know at any time about the current location and estimated arrival time of their sensitive cargo, but also wish to monitor the cargo complete parameter status during transportation. To address such a complex scenario, cargo operators not only need to employ *C-ITS* at their *logistic hubs*, they need to utilize *Intelligent Transport Units* (*iTU*) designed to enable appropriate monitoring and routing of valuable cargo throughout the *PhI*.

This chapter focuses on the design of an intelligent, safe and secure transport unit for the PhI—the iTU. The core problem of *transport chain automation* is the ability to recognize and pass the transport units on with minimum overhead while, at the same time, maintaining or increasing transport transparency, safety and security.

R. Gumzej, *Intelligent Logistics Systems for Smart Cities and Communities*,
Lecture Notes in Intelligent Transportation and Infrastructure,
https://doi.org/10.1007/978-3-030-81203-4_19

With the goal of speeding up logistics processes and reducing costs, while at the same time fulfilling the demands of the aforementioned Logistics 4.0 processes, here, the iTU together with the procedures of automated authentication and authorization of transport units and their consignors is introduced, as presented in the sequel.

19.2 State of Technology

The number of shipments with high added value has grown substantially during the past years. High value-to-weight manufactured products such as micro-electronics, pharmaceuticals, live tissue, medical devices and aerospace components are traditionally being transported as air cargo. Considering the advent of *Augmented Logistics* (*AL*) and *PhI* new and highly automated means of transportation are being introduced. Hence it is necessary to sum-up best practices and form new ones in relation to the new technologies and services.

19.2.1 Cargo Strategy

According to the IATA Cargo Strategy [2], the ten industry key priorities by 2020 were:

1. enhancing safety,
2. improving security,
3. pushing for smarter regulations,
4. strengthening the value proposition of air cargo,
5. driving efficiency through global standards,
6. modernizing air cargo,
7. improving quality,
8. protecting cash,
9. strengthening partnerships, and
10. building sustainability.

Safety is the first priority of the IATA Cargo Strategy. Some commodities may endanger the safety of carriers, their passengers and/or crews, if not shipped in accordance with stipulated regulations. With respect to these concerns, fire-resistant *Unit Load Devices* (*ULD*) have been designed which are also used to maintain the contained cargo within a specified temperature range.

As part of IATA's special cargo handling and temperature control regulations [3], a new approach to transporting pharmaceuticals was introduced, using *logistic cold chain* and *Logistics 4.0* concepts to improve the existing IATA Cargo Strategy.

The IATA *special cargo handling regulations* mostly relate to transporting live animals, perishables and pharmaceuticals. Transporting healthcare products by air requires most complex logistic processes, specific equipment, storage facilities and harmonized handling procedures to maintain shipment integrity. All special cargo requires regulations and standards for documentation, handling and personnel training.

The *Temperature Control Regulations* (*TCR*) are a comprehensive guideline, designed to enable stakeholders involved in transport and handling of pharmaceutical products to safely meet the temperature control requirements. TCR contain the requirements and standards for transportation and handling of time- and temperature-sensitive healthcare products, considering pharmaceutical product information based on WHO guidelines. Moreover, TCR provide access to the most current and efficient practices for pharmaceutical operations by guaranteeing shipment's compliance with international and/or local regulations. TCR include:

- up-to-date airline and government requirements pertaining to the transport of healthcare and pharmaceutical products,
- requirements for handling, marking and labeling,
- necessary packaging requirements,
- information on handling procedures, and
- documentation needed when transporting healthcare products.

The industry often requires pre-set environmental parameters to be maintained throughout the entire transport chain. When transporting sensitive goods, intermodal transport units must enable environmental parameters monitoring to be able to maintain them within thresholds. Upon arrival at a distribution point, any unit should be checked, according to the protocols prescribed by the TCR. Additionally, they should be checked by examining the container for defects and the seals on its openings for tampering.

Being equally important, security measures need to be both efficient and effective to be sustained. To be effective, the operations of shipping companies need to be transparent, reliable and predictable. To be efficient, their compliance with security regulations needs to be maintained by the industry itself in order to be well integrated and to sustain transportation efficiency. Global standards are needed, e.g. to ensure transport tracking and/or 48 h end-to-end shipping time, if demanded by the customer. The vision is to have a paperless industry, and to be able to rely on high-quality data available on demand to all relevant stakeholders. To improve the service quality (*QoS*), the transportation industry needs to maintain the *dependability* and *predictability* of its services.

The above mentioned priorities pertain mainly to air transport, that typically represents less than 1% of world trade by volume. But, on the other hand, it also represents 35% of world trade by value. Considering the increasing numbers of higher-value shipments and their demand for additional safety and security, the aforementioned priorities shall pertain to most of the world's transports in the future. Hence, all future cargo strategies should consider the above priorities to ensure sustainability of the world's trade.

19.2.2 Legal Framework

Owing to legal ordinances and the increasing risk potential, transport units must be examined for explosives, especially before loading them to airplanes or ships. This process, which may even be needed multiple times throughout a single transport chain, introduces additional overhead, causing increased transport costs and slowing down logistic processes.

To prevent multiple examinations of transport units after their initial loading, the concept of an *Authorized Economic Operator* (*AEO*) has been defined in the Directive (EC) No. 648/2005 [4]. AEOs, their responsible personnel in particular, are liable for their transports and obliged to distribute their goods in compliance with the relevant security regulations of their companies as well as their supply chains. By adhering to these requirements AEO shipments can progress through customs more quickly, hereby reducing transportation costs and time while maintaining transport safety. To provide for additional security and efficiency, producers, distributors and *logistic hubs* have been joined in a common regulation by the Directive (EU) No. 185/2010 [5], introducing the concepts of a *known consignor*, *account consignor* and *regulated agent*. A *regulated agent* (*RA*) is an agent, freight forwarder or any other entity that handles cargo and ensures security controls with respect to cargo and mail. A *known consignor* (*KC*) is a consignor who originates cargo or mail for its own account and whose procedures meet common security rules and standards sufficient to allow carriage of cargo or mail on any aircraft. An *account consignor* (*AC*) is a consignor who originates cargo or mail for its own account and whose procedures meet common security rules and standards sufficient to allow carriage of that cargo or mail on all-cargo or all-mail aircraft only.

According to the information for cargo handling entities in non-EU countries [6], air carriers that fly cargo or mail from a non-EU airport to an EU airport (*ACC3s*) must ensure that all cargo and mail carried to the EU is physically screened or comes from a secure supply chain, which is validated according to the EU regulations. Any entity not being one of the beforementioned entities is an unknown entity and may not be part of a secure supply chain. All cargo or mail coming from an unknown entity needs to be screened by or on behalf of an ACC3 or by a third country regulated agent before being loaded on board an aircraft bound for the EU.

In order to to obtain their status known consignors and regulated agents need to implement an appropriate security policy, according to Directive (EU) No. 185/2010. As such they are listed and their status made visible to all cargo handling participants in a corresponding database of known consignors maintained by the European Union, to be recognized as such and their shipments to be treated accordingly. The directive also describes measures they need to apply in order to protect their own transport units as well as those ones in transit [7]. To maintain their status, these companies must uphold the prescribed security measures and allow unannounced inspections by European and national supervisors. They are reaffirmed at regular time intervals that are not more than five years apart.

Upon arrival of a transport unit, a regulated agent, air- or ship-carrier needs to check the consignor's status and handle the transport unit accordingly. Inquiries about consignor status are made to the mentioned EU database, holding data on the current status of all known consignor's locations. In case the transport unit does not originate from a known consignor, conforming to Directive (EU) No. 185/2010, the transport unit is examined in concordance with all prescribed security measures. In it is identified as a known consignor's transport unit, however, most security checks can be omitted and it can proceed faster.

The core problem of automatically authenticating and authorizing transport units shipped by known consignors is the ability to recognize and swiftly pass them on without additional overhead. Although inquiries to the EU database of known consignors can be processed via secure communication channels, the data on and at the transport units are still being transmitted unsecured. Printed one-dimensional (barcode) or two-dimensional (QR-code) identifications can easily be falsified and/or exchanged. Therefore, here a patented protocol [8] for automated authentication and authorization of transport units and their consignors has been introduced as a remedy. It was devised with the objective to make use of the established regulations and speed up logistics processes.

19.2.3 Conceptual Model

The main components of a generic *Conceptual Information Model* (*CIM*) for integrated freight and fleet management within the PhI are the following.

Physical Internet (*PhI*)—an open global logistic system based on physical, digital and operational interconnectivity as achieved by encapsulation, interfaces and protocols [9]. The Physical Internet does not manipulate physical goods directly, but exclusively *PhI Containers* explicitly designed for the Physical Internet and encapsulating physical goods inside of them. It enables efficient, sustainable, adaptive and resilient Interlogistics 4.0 solutions [1].

Internet of Things (*IoT*)—semantically, IoT [10] represents the worldwide network of interconnected and uniquely addressable objects and is based on standard communication protocols. The novel paradigm is rapidly gaining ground in the scenario of wireless telecommunications. Its basic idea is the pervasive presence of a variety of things or objects such as RFID tags, sensors, actuators or vehicles around us, which are able, by unique addressing schemes, to interact with each other and to cooperate with their neighbors in order to reach common goals [11, 12]. In transport applications the Internet of Things (IoT) is sometimes considered as a synonym for PhI, however, these are two very different things. IoT is a network of devices that create and exchange data, whilst PhI is network of vehicles, routes and hubs along which transportation packages are transferred.

Cyber Physical Systems (*CPS*)—connect the real (physical) world of objects and things with the virtual (cyber) world of software and services by means of sensors, actuators and embedded computing devices. Such systems are required to consider real-world effects and contexts of informational processes [13]. CPS may be improved by engineering *self-adaptive processes* based on the principle of the so-called *MAPE-K feedback loop*, which consists of the phases monitor (M), analyze (A), plan (P) and execute (E), being repeatedly executed while sourcing data to and from a knowledge base (K). Here, the concept of MAPE-K feedback loops, as known from engineering of *self-adaptive systems* [14], is transferred to process execution.

A *Cargo Information Model* includes generation and management of digital representations of the physical and functional characteristics of cargo [15, 16]. The result of this modeling process leads to a shared knowledge resource which supports tracking and tracing of cargo from the place of origin, over to conveyance, followed by its operational life, until arriving at its destination. The resource should be used during the cargo's operational cycle—from deployment until the arrival at its destination. It provides cargo tracking information, such as physical and chemical characteristics, geometry, shape, transportation state, ownership, location etc.

Ontology—ontologies are a formal way to define the structure of knowledge for various domains with nouns representing classes of objects and verbs representing relations between objects. Ontologies resemble class hierarchies in object-oriented programming. As they are meant to represent information on the Internet coming from all sorts of heterogeneous data sources, they are expected to be rather flexible and evolving almost permanently. The *Web Ontology Language* (OWL) is a family of knowledge representation languages for authoring ontologies characterized by formal semantics. It is built upon the W3C XML standards for objects called *Resource Description Framework* (RDF). Both, OWL and RDF are used to construct and maintain the aforementioned knowledge bases (K).

Intelligent agent—intelligent agents are objects performing autonomous actions to accomplish specific tasks and are meant to perform the monitoring (M), analysis (A), planning (P) and execution (E) of MAPE-K feedback loops operated by CPSs. In the course of overlay network's implementation, model-based reflex agents (cp. [17]) should be employed, since agents cannot perceive their complete environment within a network. Initially, model-based reflex agents select their actions according to movement-related condition-action rules, which only depend on a model of the world, regardless of the current perception of the environment. Later-on they can source more information from the knowledge base (K) to perform informed decisions in their environment. The model-based reflex approach is natural for cargo tracking and tracing, because multiple threads of control naturally match the distributed and ever-changing nature of the underlying data sources affecting higher-level decision-making processes. The approach allows to more easily manage detection of and response to important time-critical stimuli, which may occur at any time from any of a large number of different sources. Hence, model-based reflex agents are considered highly beneficial for customers, clients, and various cargo agents from the point of view of keeping track of cargo from deployment to delivery. Besides, they can also assist in recognizing deviations from planned schedules and taking corrective actions through Cooperative Intelligent Transport Systems.

Cooperative Intelligent Transport System (*C-ITS*)—C-ITSs are considered ubiquitous as part of Logistics 4.0. The purpose of C-ITS is to make transport more efficient, clean, safe and cost-effective as well as to offer extensive opportunities for the development of business and innovation ideas in the transport domain. They are already used in some local and regional domains. The next big step to their global utilization is to construct the PhI, along which *PhI Containers* can be transported.

19.3 iTU Design and Protocols

19.3.1 Physical Composition

The *iTU design* comprises two main parts, viz. a container (cp. Fig. 19.1) for proper cargo storage and a smart control device (CPS) (a) charged with monitoring, control and data handling. This controller basically consists of a microcontroller, an RFID interface, a WiFi interface, various sensors for, e.g., temperature, humidity, pressure, acceleration or orientation, and tamper switches (cp. Fig. 19.5).

The sensors within the unit are attached to the controller to allow for *Remote Condition Monitoring*, which represents one of the most important services having a fundamentally positive impact on the quality of service of handling PhI containers. The ability to access cargo status information in real time is essential for support services, especially since it enables more efficient root cause analysis and solution development.

To ensure low power consumption, upon positive RFID identification WiFi radio would be turned on for data interchange, being automatically turned off again as soon as the necessary communication protocols are through.

A host of measures is foreseen to effectively protect the iTU and its controller both mechanically as well as against hacking attacks and malware intrusion.

19.3.2 Information Model

The *Conceptual Information Model* of the *iTU* (cp. Fig. 19.2) corresponds with the guidelines on Transport Regulatory Uses of Telematics in Europe [18].

Cargo-related information can be divided into two main groups and several categories according to different standards. The two main groups represent static and dynamic information:

1. Static information about cargo includes cargo identification, date, type, quantity, value of cargo and owner, which are stored in the cargo database.
2. Dynamic information about cargo includes cargo temperature, humidity, routing, status etc., which can initially be stored in the iTU's controller and the cargo database for comparison.

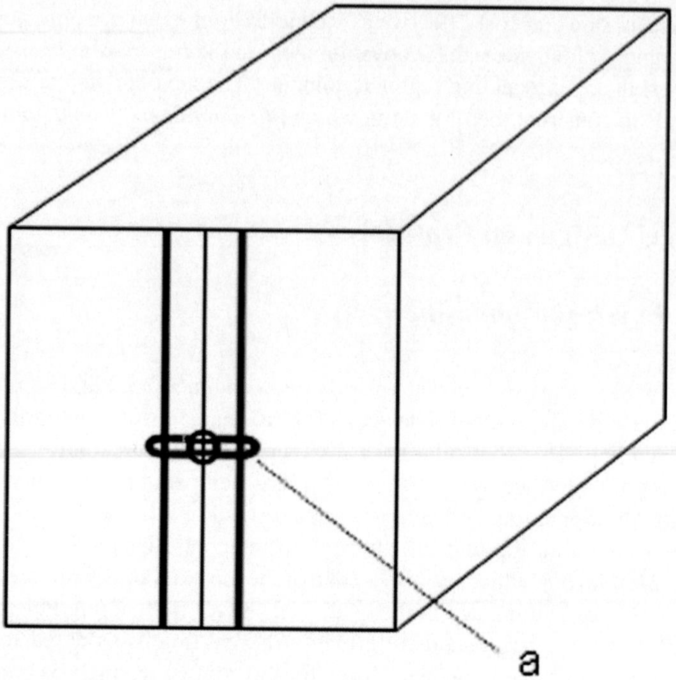

Fig. 19.1 Intelligent transport unit with **a** seal

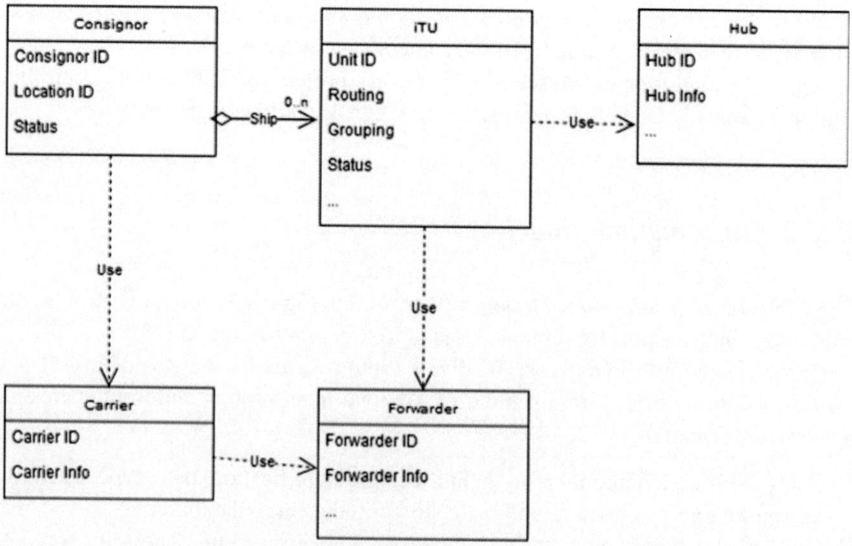

Fig. 19.2 Conceptual information model of the iTU

In case by its sensors the iTU detects value changes of dynamic information by comparing current read-outs with the initially saved status information, it communicates the changes to the back-end C-ITS system. Besides the aforementioned telemetry data, the cargo database also holds indirect cargo information, including data on the transport order, cargo owner, transportation company and repository.

19.3.3 Functional Model

The beforementioned *intelligent agent* (*IA*) technology can be used to handle transportation networks in their entirety. In the process of cargo tracking, all stakeholders as well as the cargo units in a cargo conveyance themselves can be represented by IAs ("*digital twins*"). If a problem occurs during a conveyance process, the agents can negotiate a plan for problem solving. All agents cooperating in a cargo conveyance transaction constitute an *iLS* overlay network within the *PhI*. Model-based reflex agents are applied to solve problems that are difficult or impossible to handle by any individual agent alone.

Figure 19.3 presents the collaboration among participants within an iLS network in a typical scenario of *iTU routing* from shipment to delivery. Interaction protocols are defined according to the Consignors', Carriers', Forwarders' and logistic Hubs' standard procedures.

A large shipment may first be broken down to a number of iTUs. They are assigned their origin and destination. Before they are dispatched, their status information is initialized (location and cargo state determined by sensors).

After an iTU is dispatched, it begins its journey towards its destination—from one logistic hub to another in a fastest and most efficient manner. At every logistic hub an iTU's current status and location are automatically logged and forwarded to its forwarder, carrier and consignor. Upon telemetry request, its current status may also be returned to the interested party.

The cargo conveyance procedures are enclosed by a *MAPE-K* condition feed-back loop. While the protocols are performed, all pertaining participants in a transport chain are being acquainted with an iTU's status and can act accordingly, e.g. in case a monitored status variable is above or below a given threshold.

If all information obtained from the protocol sequences, viz. identification, classification, health monitoring, routing and grouping, yield values corresponding with prescribed rules and thresholds, no action is needed. Otherwise, corresponding actions are requested by appropriate signals through the cargo handling agents to the back-end *C-ITS* systems.

Upon reaching its final destination, an iTU announces its arrival to its forwarder, carrier and consignor. Once all iTUs of a shipment have arrived, the consignor confirms the fulfillment of the shipping assignment, and the carrier confirms the fulfillment of the shipping orders. Hereby the cargo conveyance transaction through the PhI is complete.

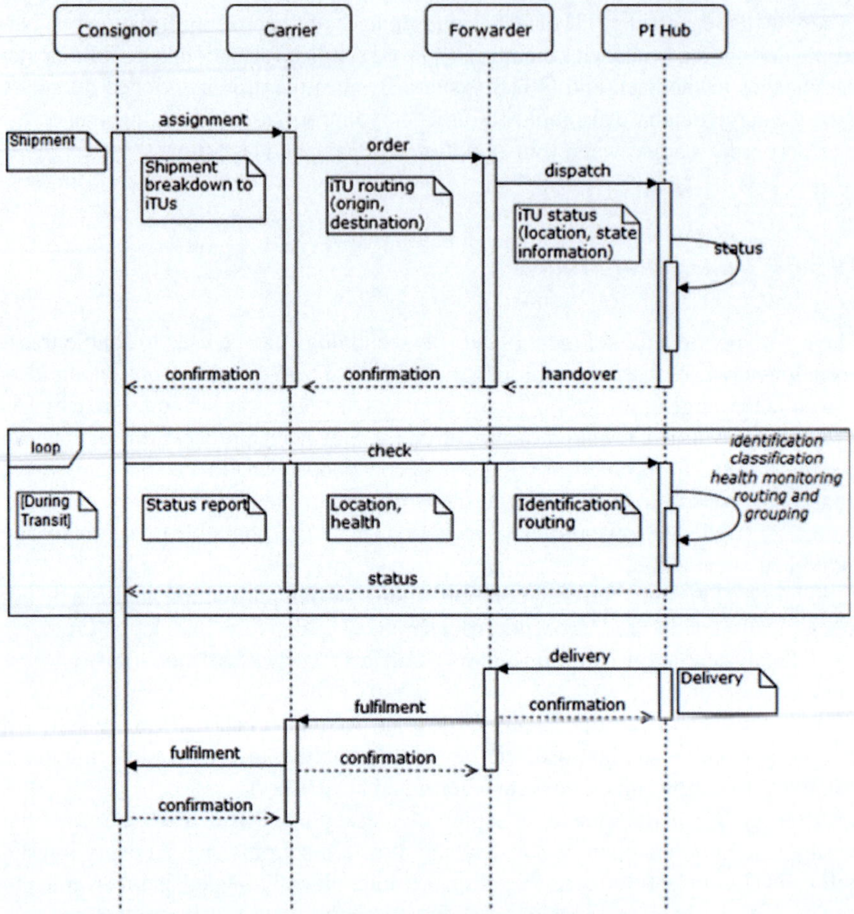

Fig. 19.3 Interaction protocol between the iTU and its handling agents

19.4 iTU Safety and Security

A protocol [8] for automated authentication and authorization of transport units and their consignors was developed with the objective to speed up logistics processes and to reduce costs while raising transport security to the level defined by Directive (EU) No. 185/2010. In response to the core problem of reducing the overhead while identifying the transport units and determining the status of their consignors, the presented automated authentication and authorization procedure for known consignors' transport units has been introduced.

19.4.1 Automated Authentication and Authorization

The authentication and authorization procedure of known consignors iTUs has been designed in a way that guarantees transport unit's safety as well as confidentiality of the data being transmitted between a regulated agent (distribution center) and the unit, the distribution center and the authorization authority of known consignors as well as between the distribution center (regulated agent) and known consignors. It has been defined as a protocol (cp. Fig. 19.4) and is composed of the following steps:

1. While preparing a transport unit for shipment, a known consignor seals the unit and places a tag on its opening ((a) in Fig. 19.1) in a way which prevents it from being opened without removing the tag. In the RFID tag's chip (cp. Fig. 19.5) the transport unit's identification (ID), declaration and authorization bit string (ABN) are stored.
2. Before a known consignor deploys a transport unit, it announces this to a distribution center by transmitting the transport unit's identification (ID) and declaration to the center.

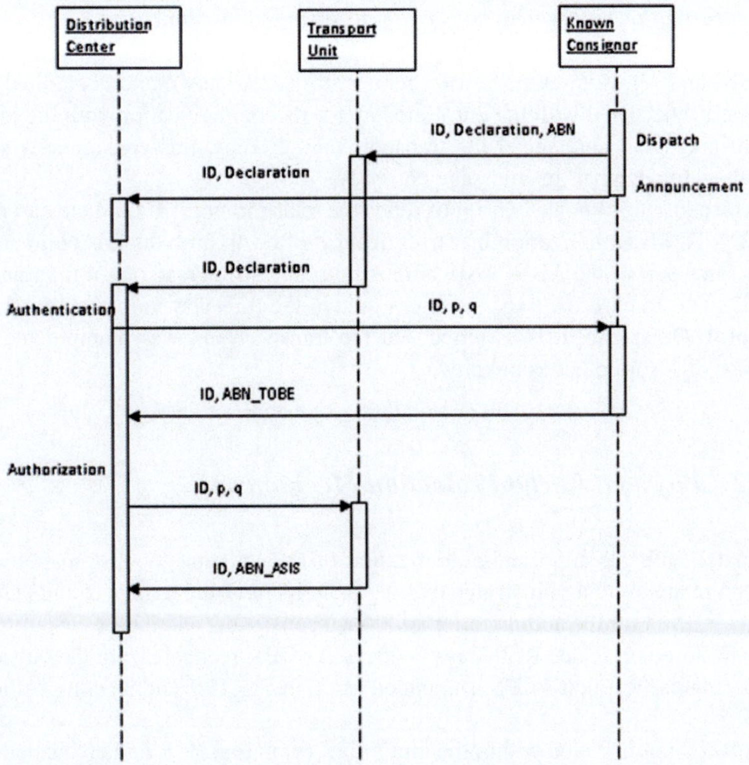

Fig. 19.4 Automated iTU authentication and authorization protocol

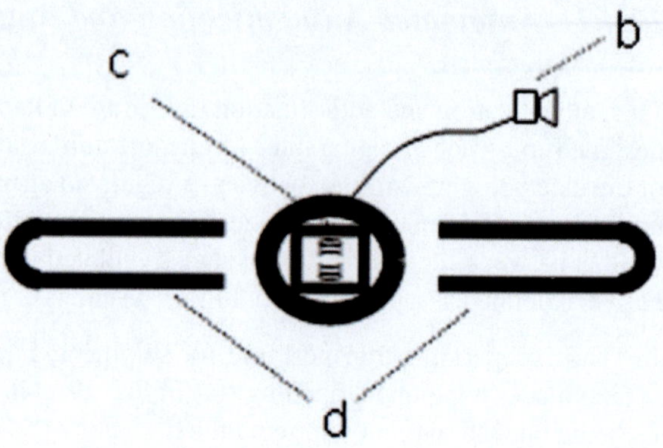

Fig. 19.5 Transport unit seal (**a**) as CPS with sensors (**b**, **c**, **d**)

3. Upon arrival of the transport unit at a distribution center, it is authenticated: it is weighed, its identification and declaration are read out from the unit's tag; the transport unit is sorted out as unknown in case it was not announced or its declaration data are inconsistent with its previously collected data.
4. In case the authentication of the transport unit and its consignor were successful, the distribution center generates two integers, random numbers $p < q$ from the interval $(0 : length)$, where $length$ matches the number of bits, constituting the ABN, and sends an authorization request to all its known and accredited consignors, encrypted with the one-time key for its communication with the known consignor who announced the transport unit. Hereby, this consignor is solely enabled to decipher this message correctly.
5. As response this known consignor sends the transport unit's ID and the part of the ABN_TOBE authorization bit string from pth to qth bit, which is compared to the same part of the ABN_ASIS authorization bit string read out of the transport unit's tag. In case the two sequences match, the transport unit is authorized as known. Otherwise, it is assumed that the transport unit was manipulated and, hence, it is sorted out as unknown.

19.4.2 Physical Cargo Protection Mechanisms

Automated authentication and authorization of known consignor's transport units relies on relatively long bit strings that uniquely identify the transport units (ABN) and can be used for the authorization of known consignors and their transport units. They are stored on sealed RFID tags inside unit's CPS, readable from the outside of transport units. The unit's CPS is depicted as (a) in Fig. 19.1 and its composition in Fig. 19.5.

An RFID tag is sealed within the unit's CPS (a) in Fig. 19.1, and connected with three sensors (c), (b) and (d) in Fig. 19.5 to the iTU, triggering an alarm in the unit's CPS upon electronic tampering, its forced removal or forced opening of the transport unit. An alarm in the unit's CPS results in the re-initialization of its authorization bit string (ABN), hereby disabling the unit's authorization and rendering it unknown.

The sensor (d) in Fig. 19.5 represents a tamper switch, being activated upon forced removal of any of the bonds, by which the CPS ((a) in Fig. 19.1) is attached to the transport unit's opening. The sensor (c) in Fig. 19.5 is activated in case of forced removal of the tag by breaking it or by electronic tampering. Any non-protocolled reading and re-writing of the data in the CPS's RFID chip would also trigger an alarm.

The sensor (b) in Fig. 19.5 is placed inside of the transport unit and connected with the tag from there. This sensor is activated upon detection of a break-in into the transport unit by creating another opening or upon breaking its connection with the CPS. Hence, the sensor (b) in Fig. 19.5 should preferably be realized as an ultrasonic sensor.

19.4.3 Cargo Data Protection Mechanisms

Encryption algorithms usually utilize the same key over a longer period of time. Hence, they are not considered entirely secure. Only encryption algorithms, for which holds that the available computing power is not sufficient to test all possible encryption keys for decrypting an intercepted encrypted message, are considered secure. This is only possible, if during encryption all possible encoded messages are equally possible, which makes any conclusions on the original message, made based on an intercepted encoded message, impossible. Corresponding with the Shannon's theorem on coding data sources [19], an encryption system is completely secure, if and only if the number of possible keys is at least as great as the number of possible messages.

To protect data transfers among regulated agents (distribution centers) and known consignors, *one-time key encryption* [20] is used, based on keys of message length. A *synchronous data transfer protocol* (cp. Fig. 19.4) among pairs of communicating parties ensures that every key is used just once and there is always a new key available. To protect data transfers among regulated agents and transport units secure RFID identification and communication protocols are used (cp. Chap. 14). Hereby, confidentiality, integrity and availability of cargo and its data is assured along the entire supply chain and the implemented integral security policy can rely on it.

19.5 Discussion

The key advantages of automatic authentication and authorization of known consignor's transport units are:

- The transport units can undeniably and justifiably be associated with known consignors.
- The transport unit's security throughout supply chains is ensured.
- Falsifications of transport units are recognized automatically.
- Due to automated procedures for authentication and authorization of transport units, less or even no personnel, who might manipulate the units, may be permitted in their surroundings.
- Safety and security of the data flow, associated with transport units, is guaranteed.

In conformance with the procedure for automated authentication and authorization of known consignor's transport units all telecommunications concerning the transport units and their transport processes fulfill the *general goals of security*:

- *confidentiality*, i.e. security of access,
- *data integrity*, i.e. protection from unauthorized data manipulation,
- *authenticity*, i.e. protection from forgery, as well as
- *responsibility*, i.e. indisputability.

The combined use of the described iTUs with their associated procedures and automated authentication and authorization protocols implemented can guarantee the safety of transport units and the security of their associated data flows in conformance with the Directive (EU) No. 185/2010, dated 4. March 2010. In addition, the overhead associated with the handling of such transport units is minimized.

19.6 Conclusion

To be able to handle air shipments with high added value by *C-ITS*, a special kind of freight containers appropriate for air cargo was designed, which can maintain prescribed micro-climatic conditions inside of them as well as communicate their parameters to their consignors. The safe and secure intelligent transport unit (*iTU*) has been designed in correspondence with the International Air Transport Association's (IATA) standards for transporting pharmaceuticals according to the principles of the *Internet of Things* and the arising *Physical Internet*. The most important improvement to be brought about by iTUs in contrast to existing solutions is the concept of smart shipping containers.

The described approach to automated authentication and authorization of consignors and their consignments within secure supply chains reduces the number of persons, involved in goods manipulation, the costs of their training and certification, as well as the risk of tampering. At the same time, it provides for automatic authentication and authorization of transport units and their consignors and secures the information flow accompanying the shipments, which has already proven beneficial in maritime transport. Along with these benefits it introduces only a minor overhead concerning the implementation of the iTU smart control devices and information interchange protocols. The proposed authentication and authorization procedure for known consignors and their transport units shall, hence, minimize the associated overhead and ensure the confidentiality of the information flow among known consignors and regulated agents with high reliability. In the long run its implementation should result in a spread of secure supply chains including also cross-docking options with multimodal transport (e.g.: plane, ship, train, truck combinations).

Considering their implementation, iTUs are a suitable form of *PhI Containers*, to be managed by C-ITS throughout transport overlay networks of the *PhI*. As part of the envisaged PhI transportation processes, also the concept of *Structural Health Monitoring* [21] for handling environmental and operational damages on smart shipping containers should be adopted with the final goal of autonomous process adaptation among C-ITSs to deal with different extraordinary scenarios. By this approach, the unit's autonomy and resilience would increase throughout the conveyance processes due to its *self-adaptation* and *self-healing* capabilities. By employing process metamodels and execution scenarios while implementing feedback loops, the proposed solution would eventually evolve to the complete *PhI OSI model*.

Last but not least, according to *IATA Cargo Strategy*, iTUs, C-ITS, PhI and iLS combine regulations with actions to proactively support the functional areas of the future *IATA Cargo Delivery Model*: Safety, Special Cargo, Border Management, e-Cargo & Quality, Operations and Industry Management [2].

References

1. Montreuil B, Meller RD, Ballot E (2013) Physical internet foundations. Springer, Berlin, pp 151–166. https://doi.org/10.1007/978-3-642-35852-4_10
2. IATA: Iata cargo strategy. https://www.iata.org/whatwedo/cargo/Documents/cargo-strategy.pdf
3. IATA: Temperature control regulations (tcr). https://www.iata.org/en/publications/store/temperature-control-regulations/
4. EU: Regulation (ec) no 648/2005 of the european parliament and the council of 13 april 2005 amending council regulation (eec) no 2913/92 establishing the community customs code. https://eur-lex.europa.eu/legal-content/EN/TXT/HTML/?uri=CELEX:32005R0648
5. EU: Commission regulation (eu) no 185/2010 of 4 march 2010 laying down detailed measures for the implementation of the common basic standards on aviation security. https://eur-lex.europa.eu/legal-content/EN/TXT/HTML/?uri=CELEX:32010R0185
6. EC: Information for cargo handling entities in non-eu countries. http://ec.europa.eu/transport/modes/air/security/cargo-mail/entities_en.htm
7. Aviation Authority: Guidance for known consignors. https://www.iaa.ie/docs/default-source/misc/guidance-for-known-consignors.pdf
8. Gumzej R, Halang WA (2017) Automatisierte authentifizierung und autorisierung von transporteinheiten bekannter versender
9. Montreuil B (2011) Toward a physical internet: meeting the global logistics sustainability grand challenge. Log Res 3:71–87. https://doi.org/10.1007/s12159-011-0045-x
10. ITU: Global information infrastructure, internet protocol aspects, next-generation networks, internet of things and smart cities. https://www.itu.int/rec/T-REC-Y/en
11. Atzori L, Iera A, Morabito G (2010) The internet of things: a survey. Comput Netw 54(15):2787–2805
12. Xiao L, Wang Z (2011) Internet of things: a new application for intelligent traffic monitoring system. JNW 6:887–894. https://doi.org/10.4304/jnw.6.6.887-894

13. Brun Y, Di Marzo Serugendo G, Gacek C, Giese H, Kienle H, Litoiu M, Müller H, Pezzè M, Shaw M (2009) Engineering self-adaptive systems through feedback loops, pp 48–70. Springer, Berlin. https://doi.org/10.1007/978-3-642-02161-9_3
14. Seiger R, Huber S, Heisig P, Assmann U (2016) Enabling self-adaptive workflows for cyber-physical systems, pp 3–17. https://doi.org/10.1007/978-3-319-39429-9_1
15. Backus J (1959) The syntax and semantics of the proposed international algebraic language of the zurich acm-gamm conference. In: IFIP congress
16. Wombacher A (2011) How physical objects and business workflows can be correlated, pp 226–233. https://doi.org/10.1109/SCC.2011.24
17. Gasevic D, Djuric D, Devedzic V (2006). Model driven architecture and ontology development. https://doi.org/10.1007/3-540-32182-9
18. Transport Certification Australia Limited (TCA): Transport regulatory uses of telematics in Europe. https://tca.gov.au/wp-content/uploads/2020/01/Transport-Regulatory-Uses-of-Telematics-in-Europe-Report-Vol-1-general-release-web.pdf
19. Shannon CE (1948) A mathematical theory of communication. 27:379–423, 623–656
20. Halang WA, Komkhao M, Sodsee S (2014) Secure cloud computing. In: Boonkrong S, Unger H, Meesad P (eds) Recent advances in information and communication technology. Springer International Publishing, Cham, pp 305–314
21. Abdelgawad A, Mahmud MA, Yelamarthi K (2016) Butterworth filter application for structural health monitoring. Int. J. Handheld Comput. Res. 7:15–29. https://doi.org/10.4018/IJHCR.2016100102

Chapter 20
Use Case: Smart Mobility

Problem 20.1 By 2050, 67% of the world population is expected to live in cities and other urbanized areas. In more developed regions, the expected share is even higher, up to 86% [1]. Rapid urban growth is causing the increase in traffic congestion that affects many aspects of our lives. Travel delays and road accidents due to traffic congestions, stress levels of drivers, increasing air pollution and greenhouse gas emissions, as well as huge transportation costs, are some of the most significant negative consequences of traffic increase [2, 3]. The transportation sector plays an important role in air pollution and noise emissions and is the main contributor to anthropogenic pollutant emissions [4]. Up to 88% of total emissions are caused by road transport [5]. As a result, traditional transport systems will no longer be able to ensure road safety, meet emission thresholds, and prevent their increasingly negative impact on the environment [6].

Solution 20.1 The right answer to the problems associated with the growing population in cities could be the concept of a *Smart City* [7]. The concept is closely related to traffic, as one of the objectives of implementing a smart city is to improve the traffic situation in cities. Therefore, *smart transportation* or, more precisely, *Intelligent Transport Systems* are among the main components of any smart city. A *Smart City* includes a *smart grid*, *intelligent infrastructure*, *smart energy*, *smart healthcare*, *intelligent technology* and *smart governance* [8–10]. The advantages of intelligent transportation can improve traffic situation in various ways [11] and are not limited to the cheapest, shortest or fastest route. They can also reduce greenhouse gas emissions by reducing congestions [8]. The main objective of this research [12] was to demonstrate the potential of computer simulation in the process of making controlled changes in *urban traffic management* and in the process of introducing new traffic-oriented Smart City services.

R. Gumzej, *Intelligent Logistics Systems for Smart Cities and Communities*,
Lecture Notes in Intelligent Transportation and Infrastructure,
https://doi.org/10.1007/978-3-030-81203-4_20

20.1 Introduction

A *Smart City* is considered an efficient, technologically advanced, green and socially inclusive city [13]. *Sustainability* and *ICT-support* are the most common additional attributes usually associated with smart cities [1]. In contrast to traditional cities, smart cities are safer, faster, friendlier and above all—greener [8–10]. As outlined in Chap. 9, due to ICT technology associated with the emerging smart cities, more and more data is being collected. This data provides a knowledge base for the analysis of urban ecosystems [14].

According to Geotab [15], smart cities are looking for ways to reduce the negative consequences of traffic associated with transportation of people and goods. Since in cities most traffic is associated with people, the term smart mobility is often used instead of smart transportation [15]. Thus, a very simple and cost-effective organizational improvement is to encourage people to share vehicles, use public transport, walk or cycle to their destination [16]. Another, more technical solution is to use smart city infrastructure to help avoid traffic congestions, being the main source of pollution.

To properly manage the smart city infrastructure, the *knowledge base*, representing the different associated entities, overlay networks, and services of a city in a digitized form, may be utilized. City traffic management mainly comprises the management of road works, parking lots, public transportation, as well as smart sensors and actuators like traffic counters, cameras, signalization, etc. Based on planned and extraordinary events different scenarios may be devised and invoked, when certain situations appear, to prevent congestions resulting in stalling and accidents.

20.2 Methodology

In order to establish an *intelligent transportation network* in a smart city, a number of small steps can be taken. To reduce traffic, it is very important to have a clear picture of the actual transport flows [17]. A consistent and coherent transportation model is necessary to improve the traffic situation in a city [18]. Hence, automated data acquisition, classification and processing are prerequisite to any reasoning on possible improvements of the traffic overlay network of a smart city. Information on all elements of the transport system at micro-level is needed, including information about transport participants [9]. It is also important to understand the changes achievable by traffic regulation in terms of traffic reduction, as positive changes on one side may have negative effects on the other side of the city. Moreover, any change in the traffic situation could cause confusion and disapproval, and even lead to road accidents with fatal consequences. Before any changes are implemented in the real world, it is important to analyze all consequences of the changes with the help of a suitable computer model [19]. A suitable way to build, maintain and use such a model is by computer-aided network simulation. *Traffic simulation* can contribute to the analysis and forecasting of traffic and human behavior in traffic [20].

Network simulation on the traffic overlay network of a city represents the basis of this research. In *network simulation*, individual entities are created and placed at an origin point and then routed through the network to their destination point, at which they leave the simulation. The simulation ends when all vehicles have reached their destinations. The forwarding and routing of simulation entities through the network is based on the entity and network properties (e.g. a pedestrian is slower than a car and may use the road only at marked intersections, while a car is faster and may advance on dedicated connections only). In addition, any traffic congestions may cause an entity to stall and/or be rerouted to a less occupied part of the network for load balancing.

Network throughput and *entity lifespan* are the two critical performance indicators that determine its quality of service—where the first one is being maximized and the second one minimized for optimal performance. In this research the negative effects of road closures on the negative side and the introduction of smart city solutions to reduce traffic and emissions on the positive side have been considered [12]. The gap between performance- and emission-based comparisons of different traffic situations shall be overcome as indicated in the envisaged implementation of the *smart city solution*.

20.2.1 Simulation

This research was founded on the microscopic road traffic simulation environment SUMO (Simulation of Urban Mobility). In *microscopic road traffic simulation* each traffic participant's behavior is modeled individually. As such, SUMO is capable of road traffic simulation modeling the scale of a city [18]. It was developed for the evaluation of pure traffic engineering solutions dealing with large scale traffic management problems and is widely used due to open access [6].

In our research several road closures in the city center and their effects on the overall traffic situation have been investigated. In addition the introduction of a Park & Ride solution at the outskirts of the city has been investigated to foresee its possible effects on the overall traffic situation as well as on traffic related emissions in the city center. Two main goals have been pursued:

1. regulation of traffic by road closures, and
2. reduction of the number of cars and required parking places on one and the increase of bicycle and bus rides in the city.

Road closures can occur for many different reasons: road works, weather conditions, natural disasters, social events, accidents or traffic calming. Their negative effects to the city economy are evident, e.g. by someone's inability of travel to work, school, to do shopping, delays in deliveries, etc. On the other hand, *Park&Ride* (*P&R*) facilities in combination with other smart solutions can have a positive affect on smart city

Fig. 20.1 Map of the sample city [12]

traffic. One of the most important features of smart mobility is effective utilization of different modes of transport, which allow citizens to choose the one that best suits their needs.

20.2.2 *Traffic Overlay Network*

In SUMO the traffic network properties are determined by the *OpenStreetMap* (*OSM*) (Fig. 20.1). The SUMO OSMWebWizard can be used to create the traffic network layout and sample traffic loads of the sample city. Individual nodes and arcs may be parameterized to determine road crossings, re-routes and dedicated roads in SUMO NETCONVERT (Fig. 20.2).

20.2.3 *Load Generation and Parameterization*

Properties of entities traveling the traffic overlay network of the city are determined by their physical properties (e.g. maximum travel speed, acceleration, deceleration, size, clearance, etc.). The classification of entities are described in [21]. Vehicle speeds and emissions are based on traffic researches made by traffic management authorities [22].

Fig. 20.2 Traffic overlay network of the sample city [12]

During the construction of the simulation model in the SUMO environment's OSMWebWizard the counts parameters [23] determine, how many vehicles are generated per hour and lane-kilometer (cp. Fig. 20.3):

$$cp = No_{v/h}/S_r \tag{20.1}$$

where: cp is the count parameter, $No_{v/h}$ is the number of vehicles per hour, and S_r is road length.

Traffic simulation in SUMO is based on the Krauß [24] car following model. The main assumption is that vehicles can progress as fast as possible while maintaining their safety distance. Pedestrians differ from vehicles in many ways, however their main property is that they have absolute priority. When some pedestrian crosses the road, every vehicle is forced to stop. In the case that a pedestrian and a vehicle are on the same lane, the vehicle will do anything to avoid a collision [25].

The safe speeds of vehicles are calculated based on (2) [26]:

$$v_{safe} = v_l(t) + (g(t) - v_l(t) \cdot t_r)/(((v_l(t) + v_f(t))/2b + t_r)) \tag{20.2}$$

where v_l is the speed of the leading vehicle in queue, v_f is the speed of the following vehicle in queue, $g(t)$ is the gap between the specific vehicle and the leading one, t_r is the reaction time of a driver and b is a maximum deceleration of the vehicle.

Fig. 20.3 SUMO scenario generation with OSMWebWizard

20.2.4 Experiment Planning

First, a traffic load—a blend of cars, trucks, buses, motorcycles, bicycles and pedestrians—needs to be created, based on traffic conditions and restrictions of the investigated area [12]. To generate a realistic traffic scenario, the counts of all mentioned traffic entities were set according to a recent traffic monitoring survey. The survey rendered a typical traffic load during the critical morning rush hour (7.15 a.m.—8.45 a.m.). The average occupancy of 1.4 persons/vehicle was determined.

In Fig. 20.4 the results at monitoring spots with the numbers of cars entering (green squares) and leaving (red squares) the observed city area are outlined. Overall, 7037 people enter and 6032 leave the observed city area in the morning rush hour. While the vehicle count parameters for our simulation model were determined by the survey, the lengths of their trips were obtained from the OSM environment.

When the load generation was done, simulation runs could be conducted based on the generated traffic overlay network and load models. A simulation run ends, when all traffic participants have reached their destinations. During the simulations various simulation outputs are automatically created to be subsequently analyzed. These outputs outline the information about the number of entities in the system, the they spent traveling from their starting to their destination points, reasons for track/lane changes, lengths of queues, traveling speeds, gaps between vehicles in a queue and a selection of the most important emissions, e.g.: amounts of CO_2, CO, HC, NO_X, PM_X, noise and fuel consumption.

To verify the suitability of the SUMO simulation environment as part of the envisaged *smart city solution*, several simulation scenarios comprising different road-

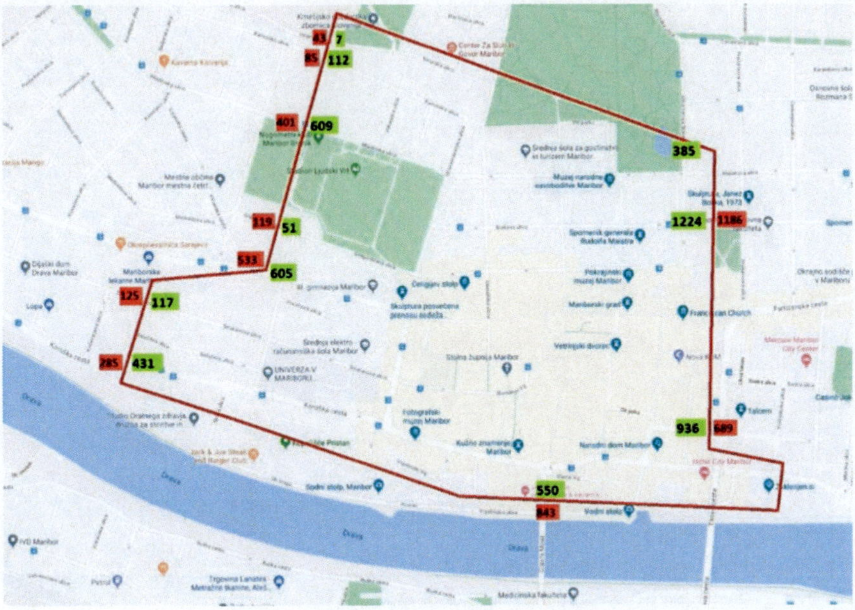

Fig. 20.4 Inbound and outbound traffic monitoring of the sampled city area [12]

closure situations have been defined and the results included in the analysis of their impact on the overall traffic situation in the city. In addition a separate scenario was defined, considering the prospective introduction of a P&R system. While road closures in cities are situations we face every day, the reasons may be very different. Anyhow, one needs to be prepared, since their consequences are similar.

In our analysis 14 different traffic scenarios have been created, each one comprising a different closure of a road section (cp. Fig. 20.5). Eventually, the individual road closure and scenarios have been joined in a single scenario in order to restrict most of the daily commuters traffic in the city center. An additional P&R scenario has been considered as a supplement to the road closure scenario. In the P&R scenario, the load distributions of the vehicles that enter the city center have been recalculated. Considering the foreseen 40% share of commuters that would use the system, the number of automobiles has been appropriately decreased while on the other hand, the numbers of buses and bicycles have been increased, to provide for their transportation [12].

Finally, the simulation outputs have been analyzed to reason on the different traffic management measures, according to the expected outcomes in terms of the established traffic performance indicators.

Fig. 20.5 Road closures in the city center [12]

20.3 Results Analysis

In simulation modeling and analysis the *representability* as well as *consistency* of simulation results need to be assured. Multiple simulation runs have been conducted for each of the defined scenarios to rule out possible deviations in the produced results, due to statistical errors. According to the observations, the reproducibility of simulation results within simulation runs could be assured. Each simulation run has been inspected to assure that every participant has been handled properly and reached its destination, considering all the imposed restrictions.

20.3.1 Average Journey Time

The most important performance indicator in traffic simulations is the time spent in a journey from start to destination. The paths of individual entities were the same in all simulation scenarios except for the Park & Ride system, due to the reduced number and structure of vehicles. Comparing the current traffic situation with the situation with road closures, an increase in travel times have been established: by about 8.8% with cars, 7.8% with bicycles and 34.5% with motorcycles. On the other hand, the travel times of trucks and buses have been shortened by 3.3% and 4.7%, respectively.

20.3.2 *Road Safety*

Traffic safety can be measured by the reasons for track/lane changes. While the numbers of lane changes may vary, considering the road closure situations, the structure of the reasons for lane changes is considered more important. As compared with the current traffic situation, the number of urgent lane changes has been increased by about 8.6% in road closure situations, while the numbers of strategic lane changes and changes due to stopping in the right lane have been slightly lowered. Overall, one could observe an increase by 5.7% in the number of lane changes with the same number of vehicles.

20.3.3 *Traffic Jams*

Another important performance indicator for intelligent traffic management are traffic congestions. As expected, the waiting times have increased in road closure situations. The highest average waiting time due to a road closure situation was determined at 74.3 s, which represents an increase by 28%, as compared with the current traffic situation.

20.3.4 *Environmental Impact*

Last but not least, the simulation results referring to the environmental impact have been observed. Fuel consumption, representing another important performance indicator, is outlined as a part of the emissions simulation outputs of SUMO. Although the fuel consumption of trucks has been 3% lower with road closures, as opposed to the current traffic situation, generally an increase by about 2% has been observed for all other types of vehicles.

The values of most pollutants in the observed area have decreased due to road closures, but their decrease is somewhat ambiguous. As expected, a significant increase of CO and PM_X emissions could be observed in road closure situations. The share of PM_X emissions is the highest with trucks, which is even more evident in the Park & Ride scenario. On the other hand, it was interesting to discover that, even though the number of buses in the P&R situation was higher, their share of emissions was lower than in the current traffic situation.

Hence, although one would expect a general improvement of air quality in the city center due to combined road closures and the introduction of the P&R solution, based on the results of the simulations, the improvement is not univocal.

20.4 Discussion

Observing the results of the analysis for the Park & Ride situation, the biggest travel time savings are achievable for automobiles, by 27.3% and the smallest for motorcycles, by 19.1%. Average journey times can also been shortened by 23% for trucks, by 26.1% for buses and by 20.4% for bicycles.

Considering road safety, by the introduction of the Park & Ride system the number of lane changes has been reduced by 58.2%, as compared to the current traffic situation. However, this could also be the consequence of a smaller total number of vehicles in this traffic simulation scenario.

About 29.7% of the waiting times due to queuing have been saved in the Park & Ride system scenario. In this case the lengths from the intersection to the last vehicle in a queue have also been shortened by about 42.4%. Moreover, the lengths of queues with the last vehicle traveling at speed lower than 5 km/h has been shortened by 48.6%.

As for the emissions, in the Park & Ride situation, the fuel consumption has been lowered by 66.2% as opposed to the current traffic situation. The biggest savings have been achieved with buses, namely by 72.0%, although their number has been increased. A significant decrease in emissions could also be observed, with the CO_2 decrease by 66.2%, CO decrease by 73.5%, HC decrease by 76.6%, NO_X decrease by 61.8% and PM_X decrease by 61.5%.

Based on the achieved consistency and coherency of the simulation models and repeatable results of the simulation runs, they are considered representative. Hence, the conclusion was that they are suitable as a decision support tool in a smart city solution.

20.5 Conclusion

One may consider the results of this research of great importance for *urban traffic planning* as well as decision-making with construction work planning. The associated models and tools can also be considered a basis for further research in the field of smart city infrastructure, as they provide for standard and open interfaces to other overlay networks of the city infrastructure (e.g. power supply, water supply, heat-supply, sewer system, etc.).

The presented simulation environment can be used as a *Decision Support* Tool for in-a-loop (*MAPE-K*) decision making—planning regulatory measures, implementing them in the model to foresee their effects, analyzing simulation results to determine, whether they achieved their goals, and acting accordingly. The smart city infrastructure's *real-time data* on traffic and pollution measurements should be included in the process to allow for more precise decision making.

By this research, the assumption that road closures effect the traffic situation in aggravation of all analyzed performance indicators has been proven. On the other hand, the proposed Park&Ride solution, has proven to deliver most of the anticipated benefits to the traffic situation in the city center. The most important finding, however, lies in the importance of simulations, being carried out beforehand to any interventions in the city's traffic situation. For example, an assumption was made that the closure of a certain road in the city center would have a positive effect on the quality of life of city dwellers of a certain area. But after analyzing the simulation results, one could conclude that even though the situation improved in the observed area, there were negative consequences in other urban areas and the overall traffic situation. Simulation results have also shown a significant increase in PM_X and CO_2 emissions due to this road closure. Hence, in order to improve citizens' quality of life, simulations of regulatory measures should be conducted to determine their overall effect.

This solution has been tested in several Slovenian cities in order to devise traffic solutions that would bring most benefits to the communities. The subsequent integration of traffic data acquisition, simulation, and results analysis on a larger scale could render improvements and mobility solutions that are yet to be discovered. Improvements in city management by joining and managing different layers of city infrastructure on a common platform are expected.

One could consider this use case a pilot implementation of a smart city's *traffic management solution* in Slovenia and a good starting point for the introduction of good practices and information interchange with related smart cities and communities.

References

1. Höjer M, Wangel J (2015) Smart sustainable cities: definition and challenges. ICT innovations for sustainability, Advances in intelligent systems and computing, pp 333–349
2. Eluru N (2010) Developing advanced econometric frameworks for modeling multidimensional choices : an application to integrated land-use activity based model framework
3. Sekhar RC (2014) Mode choice analysis: the data, the models and future ahead. Int J Traffic Trans Eng 4:269–285. https://doi.org/10.7708/ijtte.2014.4(3).03
4. Baldasano JM, Gonçalves M, Soret A, Jiménez-Guerrero P (2010) Air pollution impacts of speed limitation measures in large cities: the need for improving traffic data in a metropolitan area. Atmos Environ 44(25):2997–3006
5. Stanley JK, Hensher DA, Loader C (2011) Road transport and climate change: stepping off the greenhouse gas. Trans Res Part A: Policy Pract 45(10):1020–1030
6. Wang T, Hussain A, Bhutta MNM, Cao Y (2019) Enabling bidirectional traffic mobility for its simulation in smart city environments. Fut Gener Comput Syst 92:342–356
7. Hu B, Yixiong F, Sun J, Gao Y, Tan J (2019) Driving preference analysis and electricity pricing strategy comparison for electric vehicles in smart city. Inf Sci 504. https://doi.org/10.1016/j.ins.2019.07.039
8. Mohanty S (2016) Everything you wanted to know about smart cities. IEEE Consum Electron Mag 5:60–70. https://doi.org/10.1109/MCE.2016.2556879
9. Xiong Z, Sheng H, Rong W, Cooper D (2012) Intelligent transportation systems for smart cities: a progress review. Sci China Inf Sci 55. https://doi.org/10.1007/s11432-012-4725-1
10. Zanella A, Bui N, Castellani A, Vangelista L, Zorzi M (2012) Internet of things for smart cities. Int Things J, IEEE 1. https://doi.org/10.1109/JIOT.2014.2306328

11. Yin Y, Lam W, Ieda H (2004) New technology and the modeling of risk-taking behavior in congested road networks. Transp Res Part C: Emer Technol 12:171–192. https://doi.org/10.1016/j.trc.2004.07.009
12. Šinko S, Gumzej R (2021) Towards smart traffic planning by traffic simulation on microscopic level 11:1–17. https://doi.org/10.4018/IJAL.2021010101
13. Yigitcanlar T, Kamruzzaman M (2018) Does smart city policy lead to sustainability of cities? Land Use Policy 73:49–58
14. Nagy A, Simon V (2018) Survey on traffic prediction in smart cities. Perv Mobile Comput 50. https://doi.org/10.1016/j.pmcj.2018.07.004
15. Geotab: What is smart mobility? https://www.geotab.com/blog/what-is-smart-mobility/
16. Lennert F, Macharis C, van Acker V, Neckermann L (2017) Smart mobility and services. https://ec.europa.eu/transparency/regexpert/index.cfm?do=groupDetail.groupDetailDoc&id=34596&no=1
17. Bieker-Walz L, Krajzewicz D, Morra AP, Michelacci C, Cartolano F (2015) Traffic simulation for all: A real world traffic scenario from the city of bologna. Lecture notes in control and information sciences, vol 13, pp 47–60. https://doi.org/10.1007/978-3-319-15024-6_4
18. Krajzewicz D (2011) Traffic simulation with SUMO - simulation of urban mobility 145:269–293. https://doi.org/10.1007/978-1-4419-6142-6_7
19. Kotusevski G, Hawick K (2009) A review of traffic simulation software. Res Lett Inf Math Sci 13
20. Fernandes P, Nunes U (2010) Platooning of autonomous vehicles with intervehicle communications in sumo traffic simulator, pp 1313–1318. https://doi.org/10.1109/ITSC.2010.5625277
21. German Aerospace Center (DLR) et al, Sumo - definition of vehicles, vehicle types, and routes. https://sumo.dlr.de/docs/Definition_of_Vehicles,_Vehicle_Types,_and_Routes.html
22. Krajzewicz D, Hartinger M, Hertkorn G, Mieth P, Ringel J, Feld C, Wagner P (2003) The "simulation of urban mobility" package: an open source traffic simulation
23. German Aerospace Center (DLR) et al, Sumo - osmwebwizard. https://sumo.dlr.de/docs/Tutorials/OSMWebWizard.html
24. Krauss S (1998) Microscopic modeling of traffic flow: investigation of collision free vehicle dynamics. Deutsches Zentrum für Luft- und Raumfahrt e.V. Dt. Zentrum für Luft- und Raumfahrt e.V., Abt. Unternehmensorganisation und -information. https://books.google.si/books?id=yIU9SwAACAAJ
25. German Aerospace Center (DLR) et al, Sumo - pedestrians. https://sumo.dlr.de/docs/Simulation/Pedestrians.html
26. Song J, Wu Y, Xu Z, Lin X (2015) Research on car-following model based on sumo. In: Proceedings of 2014 IEEE 7th international conference on advanced infocomm technology, IEEE/ICAIT 2014, pp 47–55. https://doi.org/10.1109/ICAIT.2014.7019528

Appendix A
Glossary

Glossary

- *Cyber Physical Systems (CPS)* connect the real (physical) world of objects and things with the virtual (cyber) world of software and services by means of sensors, actuators and embedded computing devices.
- *Smart device* is an electronic device, connected to the Internet, which can as such offer services and co-operate with other connected smart devices in the hyper world; collectively, smart devices and their services represent the Internet of Things (IoT).
- *Internet of Things (IoT)* represents the worldwide network of interconnected and uniquely addressable smart devices and is based on standard communication protocols.
- *Global communication networks* are realized by Public Switched Telephone Networks (PSTN), Global System for Mobile communications (GSM) and satellite communication systems.
- *Global Navigation Satellite System (GNSS)* consists of multiple constellations of navigation satellites providing for accurate location and time (UTC) information anywhere on our planet.
- *Smart sensors* are sensing devices with integrated transceivers for communication among each other and/or their control systems.
- *Simple sensors* detect (changes in) environmental conditions and report their status as read-outs to a back-end information system.
- *Proximity sensors* are sensors being able to detect the presence of nearby objects without physical contact.
- *Sensors-actuators* are able to act according to their condition or in correspondence with a triggered action from the back-end information system.
- *Nanotechnology* represents science, engineering, and technology conducted at the nanoscale, which is about 1 to 100 nanometers.

R. Gumzej, *Intelligent Logistics Systems for Smart Cities and Communities*, Lecture Notes in Intelligent Transportation and Infrastructure,
https://doi.org/10.1007/978-3-030-81203-4

- *Web 2.0 or Intelligent Web* represents a social, semantic Web for managing Web content as well as the Internet of Things. It connects the humans in the social world, information and computing devices in the cyber-world, and enabled, smart devices.
- Web-Services, Intelligent Web Agents and Ubiquitous Agent Communities are *Intelligent Web solutions*, being developed on the basis of Artificial Intelligence (AI) and advanced information technology, to provide for data classification, mining and intelligent dissemination.
- *Intelligent agent (IA)* is an autonomous entity which observes (through its sensors) and acts upon an environment (using its actuators; i.e. it is an agent) directing its activity towards achieving its goals (i.e. it is rational). Intelligent agents may also learn or use knowledge to achieve their goals.
- *Ontologies* were developed in artificial intelligence (AI) to facilitate knowledge sharing and reuse. They enable a shared and common understanding of a defined domain that can be communicated between people and application systems. Ontologies are a formal way to define the structure of knowledge for various domains with nouns representing classes of objects and verbs representing relations between objects.
- The *Web Ontology Language (OWL)* is a family of knowledge representation languages for authoring ontologies characterized by formal semantics. It is built upon the W3C XML standards for objects called Resource Description Framework (RDF).
- *Agents* interact with each other by using the *Agent Communication Language* (ACL). Agents need to be able to communicate with users, system resources and with each other to collaborate and negotiate.
- *Intelligent agents* are objects performing autonomous actions to accomplish specific tasks and are meant to perform the *monitoring (M), analysis (A), planning (P) and execution (E)* phases of *MAPE-K feedback loops* of the CPSs within the ILS.
- *Augmented reality applications* need to join real and virtual worlds, offer real-time interaction, and feature three-dimensional representation.
- *Biometrics* are technologies that automatically confirm the identity of people by comparing patterns of physical or behavioral characteristics in real time against enrolled computer records of those patterns.
- *Radio Frequency IDentification (RFID)* offers a solution to automated identification of objects where RFID tags only need to get in the proximity of a scanner/reader in order to be recognized and can be made tamper-proof, if needed.
- Smart environments, wearable computers, and ubiquitous computing in general represent the upcoming *sixth generation of computing and information technology*.
- *Telehealth, telediagnostics and telemedicine* represent new health services, based on new process models and advanced information and telecommunications solutions. *Telehealth (e-health)* is the all-encompassing term, while *telediagnostics* and *telemedicine* pertain to providing diagnostic and therapeutic treatment on a distance by the use of information and communication technology. By

telemedicine, an efficient way of equally personalized, but faster multidisciplinary specialist treatment can be introduced into clinical practice.

- *Mobile health* (m-health) service is known as the practice of medical and public health support to delivering telemedicine services through mobile devices such as mobile phones and tablets.
- *Body Sensor Network (BSN)* is a a wireless juxta-corporal sensor-net on a human body.
- *Multi-factor authentication (MFA)* is an electronic authentication method in which a computer user is granted access to a website or application only after successfully presenting two or more pieces of evidence (or factors) to an authentication mechanism: *knowledge* (something only the user knows), *possession* (something only the user has), and *inherence* (something only the user is).
- *Supply Chain (SC)* is the network of organizations that are involved, through upstream and downstream linkages, in the different processes and activities that produce value in the form of products and services in the hands of the ultimate consumer.
- *Supply Chain Management (SCM)* comprises the design, planning, execution, control, and monitoring of supply chain activities with the objective of creating net value, building a competitive infrastructure, leveraging worldwide logistics, synchronizing supply with demand, and measuring performance globally.
- A common classification of *Electronic Marketplaces* distinguishes between two basic types of *e-marketplaces*:

Abbreviation

1. horizontal e-marketplaces support cross-industry functions and processes and
2. vertical e-marketplaces support business processes along the value chain for a certain line of industry and are thus targeted to specific subject areas.

- *Supply Chain Operating Networks (SCON)* bring together trading partner connectivity with software-as-a-service (SaaS) applications.
- Two approaches to *service transformation* in the digital era can be observed for efficient, service-quality-oriented supply chain management:

Abbreviation

1. centralized e-marketplace web portals, and
2. decentralized, agent-based e-marketplaces with a yellow-page dictionary service and distributed network of temporally and spatially distributed agents of supply chain partners.

- In *Supply Chain Management (SCM)* the *Supply Chain Operations Reference* (SCOR) is a common process model, based on five distinct processes: *plan, source, make, deliver and return.*

- *Zero-latency enterprise* concept means that any change in the enterprise information system is evident in any part of the information system without noticeable delay.
- *Real-time* attributes various operations in computing or other process environments that must guarantee response times within a specified time (deadline), usually a relatively short time.
- A *real-time process* is generally one that happens in defined time intervals of maximum duration and fast enough to affect the environment in which it occurs, such as responses to input signals from a computing system.
- *Real-time ability* is an important property of processes and transactions. Most of them have some temporal restrictions, which define their validity. Eventually, this influences their timeliness, which is a service quality criterion.
- *Available-to-Promise (ATP)* is a business function that provides a response to customer order inquiries, based on resource availability. It relates to timeliness in the sense of predicting, when a resource would be available. It has been derived from the push and pull ordering schemes in SCM dealing with forward or backward scheduling to determine that time.
- *Industry 4.0* (Germany) and *Made in China 2025* (China) stand for smart production of smart products. The infrastructures for enabling the realization of these concepts are the *Intelligent Web (Web 2.0)* in general and the *Internet of Things (IoT)* in particular.
- *ANSI/ISA-95* standard is an international standard developed by the International Society of Automation for developing automated interfaces between enterprise and control systems. It has mainly been used in the realizations of *MES* systems interconnecting the ERP and process levels.
- For intra- and inter-organizational information interchange XML and JSON based *Electronic Data Interchange* (EDI) standards like the ANSI ASC X12 are being used.
- *OPC Unified Architecture (OPC UA)* is a vendor-independent communication protocol for industrial automation applications. It is based on the client-server principle and allows seamless communication from the individual sensors and actuators up to the ERP system or the cloud. OPC UA is being developed as a standard IEC 62541. The OPC UA protocol specification consists of 15 parts and is still evolving.
- Advanced Peer-to-Peer *(P2P)* collaboration and message interchange models have been developed by the *Foundation for Intelligent Physical Agents* (FIPA).
- *FIPA Agent Communication Language (FIPA ACL)* defines autonomous agent's communication and cooperation protocols to access and exchange information among IoT devices and enable their collaboration.
- *Physical Internet (PhI)* is an open global logistics system founded on physical, digital, and operational interconnectivity, an open shipping system in which modular packages are routed from source to destination through the existing transportation infrastructure.

- The PhI as a reconfigurable transportation network accommodates roaming intelligent *Intermodal Loading Units* (ILU), being characterized by their ability to autonomously navigate their way through the PhI.
- Within the PhI, *PhI Hubs* represent logistic distribution centers with capacities to store and forward ILUs in the form of smart containers.
- *Interlogistics 4.0* is to be operated by *Cooperative Intelligent Transport Systems (C-ITS)* managing *ILUs* and moving them throughout the *PhI* in the same form as data packets travel the Internet.
- *C-ITS* and *PhI* represent an infrastructure that should carry out logistics services according to the *Augmented Logistics paradigm*. In *Augmented Logistics* the location or even ownership of assets or goods is irrelevant. Important is that needed assets or goods are available, when they are asked for.
- *Intelligent Logistics Systems (iLS)* constitute innovative supply chain and logistics systems that will establish transparency for the whole production system, holistic consideration of environmental and economic aspects, systematic linking of engineering with sustainability assessment and clear views of priorities, risks, and trade-offs. iLS shall also enable integration of engineering expertise with *Carbon Footprinting* and *Life Cycle Assessment*.
- *Augmented Logistics* realized by *Intelligent Logistics Systems (iLS)* represents a multitude of digital, informational and material services and resources that are complementing physical logistics systems.
- A *Cargo Information Model (CIM)* includes generation and management of digital representations of the physical and functional characteristics of cargo. The result of this modeling process leads to a shared knowledge resource which supports tracking and tracing of cargo from the place of origin, over to conveyance, followed by its operational life, until arriving at its destination.
- *Cooperative Intelligent Transport Systems (C-ITS)* are considered ubiquitous as part of Interlogistics 4.0. The purpose of C-ITS is to make transport more efficient, clean, safe and cost-effective as well as to offer extensive opportunities for the development of business and innovation ideas in the transport domain.
- An *eco-mega-city* should not only provide for a sustainable and ecologically neutral environment to urban dwellers. It should also provide for the necessary infrastructure that would enable them to learn, play and work more efficiently – a Smart City.
- *Integrated logistics support (ILS)* is an integrated and iterative process for developing material acquisition and support strategy that optimizes functional support, leverages existing resources, and guides the system engineering process to quantify and lower life-cycle cost as well as to decrease the logistics footprint (demand for logistics), making the system easier to deploy and support.
- Integrated logistics support decisions are documented in a *Life-Cycle Sustainment Plan (LCSP)*, a support-ability strategy, or an *Integrated Logistics Support Plan (ILSP)*.
- The smallest unit of a *Smart City* is a *Smart Home*.
- The basic component of a *Smart Home* is a *smart controller device* which integrates all smart devices and provides for their universal interface, so they can

be monitored and controlled by mobile phones, tablets, computers or dedicated control devices on the home Intranet. The controller device is usually realized as a router which maintains a wireless network by which connected devices can access each other and the Internet. In order for a *smart device* to become a part of a Smart Home system, it needs to be registered with the controller.

- A *Smart Home controller application* may be realized as a dedicated back-end application or Web service application and a number of front-end applications comprising Web- and mobile applications, which may also be supplemented by voice recognition, thus building a client-server architecture.
- *Smart cities*, villages, etc. rely on smart city infrastructure to manage their resources more efficiently and provide their inhabitants with a wealth of highly customizable and accessible services. While some are mandatory (electricity, water,...) the rest are utilitarian or subscription based (cable, wireless communication networks,...).
- The *Smart City infrastructure* in the form of *smart devices and services* forms a framework of the *Information Society*.
- *Smart cities and communities* are the foundations of the future *Innovation Society*.
- The individuals in the Information Society are often referred to as *digital citizens (e-citizens)*.
- *Innovation society* in a modern company is characterized by targeted but broad information sharing among co-workers. Hereby, every co-worker receives and shares ideas being instantly communicated through Information systems to ecosystems of processes - systems of systems. In an innovation society every e-citizen is a *knowledge worker*, contributing to a company's and a community's knowledge base.

Index

R. Gumzej, *Intelligent Logistics Systems for Smart Cities and Communities*, Lecture Notes in Intelligent Transportation and Infrastructure,
https://doi.org/10.1007/978-3-030-81203-4

J

K

L

M